T0181767

Ontology of Communication

Roland Hausser

Ontology of Communication

Agent-Based Data-Driven or
Sign-Based Substitution-Driven?

 Springer

Roland Hausser
Abteilung für Computerlinguistik
Friedrich-Alexander-Universität Erlangen
Erlangen, Germany

ISBN 978-3-031-22741-7   ISBN 978-3-031-22739-4   (eBook)
https://doi.org/10.1007/978-3-031-22739-4

This Springer imprint is published by the registered company Springer Nature Switzerland AG
The registered company address is: Gewerbestrasse 11, 6330 Cham, Switzerland

Ontology of Communication

Agent-Based Data-Driven or Sign-Based Substitution-Driven?

Roland Hausser

Friedrich-Alexander-Universität Erlangen-Nürnberg (founded 1743)

河宇士

Pen name given to the author by Professor Inseok Yang
President of the Korean Society of Linguistics, Seoul 1982

Preface

The precomputational foundations of theoretical computer science in the 1930s, 40s, and 50s inherited the sign-based substitution-driven ontology from mathematics and symbolic logic. Continuing from Frege, Hilbert&Ackermann, and Russell, the work of Church, Gödel, Kleene, Post, Tarski, and Turing resulted in a rich harvest of undecidable or undecided problems, such as the Entscheidungsproblem, the $P =? NP$ problem, Gödel's incompleteness proofs, the halting problem of Turing machines, and Post's correspondence problem.

To these venerable achievements, Chomsky added the claim that natural language is undecidable as well. Based on a sign-based substitution-driven ontology, the proof[1] relies on the complexity hierarchy of *Phrase Structure Grammar* (PSG, Chomsky hierarchy) and the combination of a context-free base with a transformation component, making the system, called *Generative Grammar* (GG), recursively enumerable, i.e. undecidable.

GG derivations are all started by the same input, namely the single nonterminal S node (for **S**entence or **S**tart), working like the start button of a stand-alone algorithm. The intended output is the random generation of well-formed expressions of a natural language, using the recursive substitution of nonterminal symbols which are finally substituted by terminal symbols. Just as the motor of a car requires a skilled human driver with vision and manipulation to keep the car on the road, the stand-alone generation algorithm of GG requires a native speaker and a linguist to distinguish between grammatical and ungrammatical output.

Based on the derivation principle of computing possible substitutions, GG is "not intended" [2] as a model of the speaker-hearer. Instead, the goal is a "universal" characterization of the "human language ability." This may have been misinterpreted as the implicit promise that understanding the universal human language ability would fundamentally facilitate computational language processing, despite GG's undecidability. Today, however, after more than half a century and a tremendous international effort, the promise remains unfulfilled.

Based on the different derivation principle of computing possible continuations, the goal of DBS is a computational realization of natural language communication, defined as the automated transfer of content between cognitive agents. The input to the

---

[1] Peters, S., and R. Ritchie (1973) "On the Generative Power of Transformational Grammar," *Information and Control*, Vol. 18:483–501
[2] Chomsky 1965, p. 9.

DBS speak mode is a content and the output a language-dependent surface. Speak mode derivations are inherently unambiguous, but may provide a choice between paraphrases. The input to the hear mode is a language-dependent surface and the output a content. Hear mode derivations, in contrast, may have more than one reading (ambiguity), but each reading of an n word form surface requires exactly n-1 operation applications.

The differences in the input and the output of GG and DBS, respectively, require different derivations. The substitution-based derivations of GG are vertical top-down; there is no inherent distinction between the speak and the hear mode, and no upper limit on the number of substitution operations for the length of an *output*.

DBS, in contrast, distinguishes between the speak and the hear mode from the outset. The modes share the horizontal (left-associative) direction of derivation, but take opposite inputs and outputs. In either mode, the number of operations is a function of the length of an *input*.

Following from these structural differences, GG and DBS have orthogonal hierarchies of computational complexity. The classes of DBS are the C1, C2, C3, B, and A languages. The classes of PSG are the regular, context-free, context-sensitive, and unrestricted languages.

The C-languages of DBS are so-called because they are *constant* in that each navigation (speak mode) or concatenation (hear mode) operation may take only a finite number of primitive operations, i.e. below a grammar-dependent upper bound; the only way to raise the complexity of a C-language above linear is *recursive ambiguity*. The B languages are so-called because the number of steps in an operation is Bounded by the length of the input. The A languages are so-called because they comprise only and All recursive languages.

In summary, the claim that natural language is undecidable holds for sign-based substitution-driven GG, defined as a transformation component on top of a context-free base. Agent-based data-driven DBS, in contrast, would require recursive ambiguity for any complexity degree above linear. Recursive ambiguity, however, is absent in natural language. Consequently, the processing of natural language in DBS is of linear complexity, which is a precondition for general real time performance of a talking autonomous robot.

Status Quo

A basic choice in contemporary cognitive science is between a sign-based substitution-driven and an agent-based data-driven ontology. Most of current research is invested in the sign-based substitution-driven approach, which originated in mathematical logic (Skolem 1920, Post 1936). As an alternative, this book explores the agent-based data-driven approach of Database Semantics (AIJ 1989, TCS 1992).

Manuscript Production

The camera-ready copy was made by the author using the LaTeX software. Thanks to Josef Moser of LinuxHilfe for fixing bugs and building the name index.

Background

Database Semantics (DBS, AIJ'89) is an agent-based data-driven theory of how natural language communication works. The prototype resembles natural agents in that it assumes (i) real bodies out there in the real world (embodiment, MacWhinney 2008) and (ii) an agent-internal cognition which includes an *interface* component for elementary recognition and action, a *memory* component for storing continuous monitoring, and an *operations* component for building and processing content.

In language communication, DBS agents switch between the speak and the hear mode (turn-taking, Sacks et al. 1974, Schegloff 2007). The speak mode is driven by navigating along the semantic relations in a cognition-internal content (input) resulting in cognition-external raw data (output), e.g. sound waves or pixels, which have no meaning or grammatical properties whatsoever but may be measured by natural science. The hear mode is driven by the raw data (input) produced by the speaker resulting in cognition-internal content (output). For communication to be successful, the content encoded by the speaker into raw data and the content decoded by the hearer from those raw data must be the same (minimal requirement).

Contents are built in agent-internal cognition from the classical semantic kinds *concept, indexical,* and *name,* and connected with the classical semantic relations of *functor-argument* (Chapter 4) and *coordination* (Chapter 5). The interaction between the agent-internal cognition and the agent-external raw data is based on the computational Mechanisms of (i) type-token matching for concepts, (ii) pointing at values of the on-board orientation system for indexicals, and implicit or explicit (iii) baptism for named referents. The computational complexity of natural language communication in DBS is linear (TCS'92).

Contents

1. Introduction

For long-term incremental upscaling to be successful, the computational reconstruction of a natural mechanism must be *input-output equivalent* with the prototype, i.e., the reconstruction must take the same input and produce the same output in the same processing order as the original. Accordingly, our computational reconstruction of natural language communication uses a time-linear derivation order for the speaker's output and the hearer's input. The surfaces serving as the vehicle of content transfer from speaker to hearer are raw data, e.g., sound waves or pixels, without meaning or any grammatical properties whatsoever, but measurable by natural science.[1]

1.1 Ontology

The ontology of a field of science comprises the basic elements and relations assumed to allow a complete analysis of its phenomena. For example, the Presocratics tried to explain nature based on an ontology of fire, water, air, and earth. Today, the ontology of physics is based on a space-time continuum, protons, electrons, neutrons, quarks, neutrinos, etc.

Similarly in theories of meaning in philosophy. There was a time in which meaning was based on naming; for example, the celestial body rising in the morning and setting in the evening served as the meaning of the word sun. Then meaning became defined in terms of set-theoretic denotations in possible worlds. Which ontology is required for building the computational cognition of a talking robot?

Just as an ontology without subatomic particles is unsuitable for modern physics, an ontology of computational cognition without an agent, without a distinction between an agent-external reality and agent-internal processing, without interfaces for recognition and action, without a distinction between the speak and the hear mode, without an on-board memory (database), without an on-board orientation system (OBOS), and without an algorithm for moment-by-moment monitoring is unsuitable for the task of building a talking robot.

---

[1] Thanks to Prof. MacNeilage, director of the Phonetics Lab at UT Austin during the author's visit at the Linguistics Department as a Ph.D. student (1970–1974), for the opportunity to participate as a test person in phonetic research experiments. This instilled a permanent appreciation of raw data in language communication.

R. Hausser, *Ontology of Communication*, https://doi.org/10.1007/978-3-031-22739-4_1

1.2 Computational Cognition

The ontological requirements for computational cognition were essentially laid down in the year 1945 as the von Neumann machine (vNm): the interface component of DBS (AIJ'01) bcorresponds to the vNm input-output device, the DBS on-board database corresponds to the vNm memory, and the DBS left-associative operations algorithm corresponds to the vNm arithmetic-logic.

Designing and building the computational cognition of a talking autonomous robot is not only of interest for a wide range of practical applications, but constitutes the ultimate standard for evaluating the many competing theories of natural language in today's linguistics, language philosophy, language psychology, and computer science. It leads from the sign-based substitution-driven ontology of mathematics and symbolic logic to the new (or extended) ontology of agent-based data-driven robotics in general and DBS in particular. It also leads from Generative Grammar (GG) and its attempt to discover an innate human language ability to the effective transfer of content from the speaker to the hearer by means of raw data.

Communication is successful if the content encoded by the speaker into raw data equals the content decoded from the raw data by the hearer. DBS constructs content from the three basic content kinds of (i) concept, (ii) indexical, and (iii) name. Each has its characteristic computational mechanism: concepts use computational pattern matching based on the type-token relation, indexicals use pointing at values of the agent's on-board orientation system, and names use an explicit or implicit act of baptism which inserts a named referent as core value into a name proplet (CASM'17).

1.3 Agent-Based Data-Driven vs. Sign-Based Substitution-Driven

Most analyses of natural language in today's linguistics, philosophy, and computer science rely on a precomputational, sign-based, substitution-driven ontology. Sign-based means: no distinction between the speak and the hear mode. Substitution-driven means: using a single start symbol as input for randomly generating infinitely many different outputs, based on possible substitutions by rewrite rules. Thereby different outputs are assigned the same denotation, i.e., True or False.

However, for a functionally complete, scientific reconstruction of natural language communication, the start button is uniquely unsuitable as input to the speak mode and the truth-values are uniquely unsuitable as output of the hear mode. In DBS, propositions do not denote but *are content*, and different propositions are different contents. A content is defined as a set of proplets, i.e., order-free (which is essential for storage in and retrieval from a content-addressable on-board database). Proplets are defined as nonrecursive feature structures with ordered attributes (which is essential for efficient pattern matching). The proplets in a content are connected by the classical semantic relations of structure, i.e., functor-argument and coordination, coded by address.

The ontology of DBS is agent-based and data-driven. Agent-based means: design of a cognitive agent with (i) an interface component for converting raw data into cognitive content (recognition) and converting cognitive content into raw data (action), (ii) an on-board, content-addressable memory (database) for the storage and retrieval of content, and (iii) separate treatments of the speak- and the hear-mode. Data-driven means: (a) mapping a cognitive content as input to the speak mode into a language-dependent surface as output, and (b) mapping a surface as input to the hear mode into a cognitive content as output.

1.4 Reconciling the Hierarchical and the Linear

Content serving as input to the speak mode and as output of the hear mode is defined as a set of proplets, connected by the semantic relations of structure, coded by address:

1.4.1 CONTENT OF I saw you.

$$
\begin{bmatrix}
\text{sur: I} \\
\text{noun: } \mathbf{pro1} \\
\text{cat: s1} \\
\text{sem: sg} \\
\text{fnc: } \mathbf{see} \\
\text{mdr:} \\
\text{nc:} \\
\text{pc:} \\
\text{prn: 3}
\end{bmatrix}
\begin{bmatrix}
\text{sur: saw} \\
\text{verb: } \mathbf{see} \\
\text{cat: \#n \#a decl} \\
\text{sem: past} \\
\text{arg: } \mathbf{pro1\ pro2} \\
\text{mdr:} \\
\text{nc:} \\
\text{pc:} \\
\text{prn: 3}
\end{bmatrix}
\begin{bmatrix}
\text{sur: you} \\
\text{noun: } \mathbf{pro2} \\
\text{cat: sp2} \\
\text{sem: sg} \\
\text{fnc: } \mathbf{see} \\
\text{mdr:} \\
\text{nc:} \\
\text{pc:} \\
\text{prn: 3}
\end{bmatrix}
$$

The classical semantic relations of structure are subject/predicate, object\predicate, modifier|modified, and conjunct−conjunct. The semantic relations in 1.4.1, are subject/predicate and object\predicate, indicated by **bold face** font. In order for communication to be successful, the input content of the speak mode and the output content of the hear mode must be the same (minimal condition).

1.5 Speak Mode Converts Hierarchy Into Linear Surface

The speak mode converts the hierarchy of the input content into the linear structure of the output surface by navigating along the semantic relations of structure:

1.5.1 GRAPH ANALYSIS UNDERLYING PRODUCTION OF 1.4.1

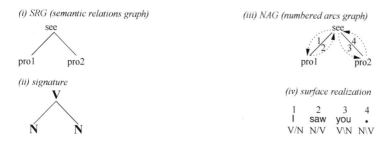

The (iv) surface realization consists of three lines, showing (1) the arc numbers, (2) the surfaces realized from the goal proplet, and (3) the traversal operations.

The operations driving the navigation in 1.5.1 are listed as follows:

1.5.2 SEQUENCE OF OPERATION NAMES AND SURFACE REALIZATIONS

arc 1:	V/N	from *see* to *pro1*	I	(TExer 2.3.8)
arc 2:	N/V	from *pro1* to *see*	saw	(TExer 2.3.9)
arc 3:	V\N	from *see* to *pro2*	you	(TExer 2.3.10)
arc 4:	N\V	from *pro2* to *see*	.	(TExer 2.3.11)

1.6 Hear Mode Re-Converts Linear Input Into Hierarchical Output

The hear mode re-converts the stream of raw input data into the hierarchical structure of 1.4.1 by incremental lexical lookup and syntactic-semantic composition.

1.6.1 GRAPHICAL HEAR MODE DERIVATION OF THE CONTENT 1.4.1

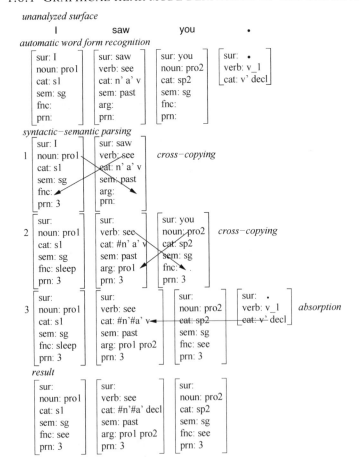

The composition is time-linear in that the current next word (lexical proplet) is related semantically to a proplet in the current sentence start (set of proplets already connected, at least partially).

The hear mode operations are of three kinds: (i) cross-copying (connective ×), (ii) absorption (connective ∪), and (iii) suspension (connective ~). Operations with the same connective may re-introduce different semantic relations of structure, for example, SBJ×PRD and OBJ×PRD, defined as follows:

1.6.2 CROSS-COPYING *pro1* AND saw WITH SBJ×PRD (LINE 1)

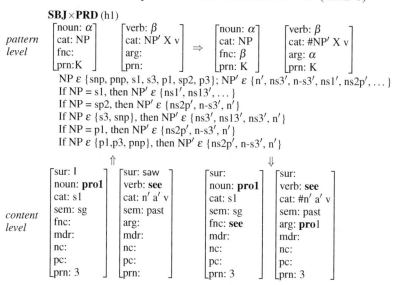

1.6.3 CROSS-COPYING saw AND *pro2* WITH PRD×OBJ^2 (LINE 2)

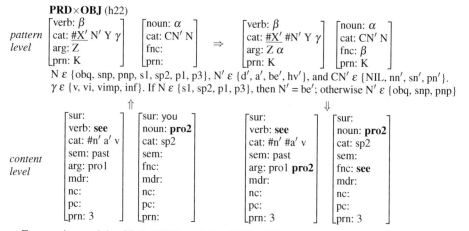

Comparison of the **SBJ×PRD** and the **OBJ×PRD** application illustrates the highly precise coding of grammatical detail, provided by the computational pattern matching of DBS. For the complete declarative analysis of I saw you. in the speak and hear mode see TExer 2.3.

1.7 Derivation Order

The regular total-order derivation of time-linear Left-Associative Grammar (LA-grammar, LAG) as the precursor of Database Semantics (DBS) differs from the irregular, partial-order derivations of Categorial Grammar (CG, bottom up) and Phrase Structure Grammar (PSG, top down):

1.7.1 THREE CONCEPTUAL DERIVATION ORDERS (FoCL 10.1.1)

| LA Grammar | C Grammar | PS Grammar |

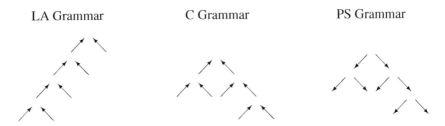

bottom-up left-associative   *bottom-up amalgamating*   *top-down expanding*

The initial empirical test of using the left-associative derivation order for the syntactic-semantic analysis of a nontrivial set of natural language expressions was programming the time-linear derivations of 221 constructions of German and 114 constructions of English during a research stay at CSLI Stanford in 1984-1986.[3]

1.8 Type Transparency

The purpose of formal grammars for fragments of natural language is (i) a linguistically well-motivated analysis of examples which is suitable (ii) for efficient automatic derivation by a computer program and (iii) for systematic upscaling. This requires *input-output equivalence* between the declarative derivation order of the formal grammar and the procedural derivation order of the parser.

Called *type transparency* by Berwick and Weinberg (1984, p. 39), input-output equivalence between a formal grammar and its parser was originally intended also in PSG:

> Miller and Chomsky's original (1963) suggestion is really that grammars be realized more or less directly as parsing algorithms. We might take this as a methodological principle. In this case we impose the condition that the logical organization of rules and structures incorporated in the grammar be mirrored rather exactly in the organization of the parsing mechanism. We will call this as follows:

1.8.1 DEFINITION OF ABSOLUTE TYPE TRANSPARENCY

- For any given language, parser and generator use the *same* formal grammar,
- apply the rules of the grammar *directly*,
- in the same *order* as the grammatical derivation,
- take the same *input* expressions as the grammar, and
- produce the same *output* expressions as the grammar.

For Phrase Structure Grammar (PSG), type-transparency is impossible. PSG is based on Post's (1936) production or rewrite systems, which were designed to mathematically characterize the notion of *effective computability* in recursion theory.[4] In this original application, a derivation order based on the substitution of signs by other signs is perfectly natural. When Chomsky (1957) 'borrowed' the Post production system under the name Phrase Structure Grammar for analyzing natural language, he inadvertently inherited the substitution-driven derivation order.

Because rewrite systems take the start symbol as input, but a parser takes terminal strings, rewrite systems and their parsers are not input-output equivalent – which means that a type transparent PSG parser can not exist. Instead, huge intermediate structures are required to reconcile the time-linear input order of the parser and the top-down substitution order of the grammar's rewrite rules (Early Parser[5], CYK Parser[6], Tomita Parser[7]).

Consequently, (i) the computational complexity of PSG is polynomial,[8] and (ii) debugging and upscaling in PSG-based parsing is greatly impeded: if a well-formed input is rejected or an ill-formed input accepted, the error must be found in the complex intermediate structures of the context-free PSG parser, which are not easy to read.

LA-grammar is not type-transparent either because the speak and the hear mode use different algorithms for different kinds of input, but the output of the speak mode is the input to the hear mode. In the hear mode, an error is located in the output close to where the time-linear derivation broke off or the ill-formed continuation began. Moreover, the error is explicitly documented in the automatic analysis serving simultaneously as the trace of the parse and the grammatical analysis.[9]

---

[3] Thanks to CSLI Stanford for their generous hospitality, especially by providing the at the time most advanced workstations by HP with a team of helpful operators, and to the DFG for a five year Heisenberg grant. The research stay was initially intended to program the Montague Grammar defined in *Surface Compositional Grammar* (SCG). Even though the syntactic-semantic λ-derivations of surfaces into formulas of intensional logic were explicitly defined to high standard, a reasonable programming of the 'fragment' presented unsurmountable difficulties. In response, a time-linear approach was developed, programmed, and published as NEWCAT, including the source code written in Lisp.

[4] See for example Church (1956), p. 52, footnote 119.

[5] Early 1970.

[6] Cocke and Schwartz 1970, Younger 1967

[7] Tomita 1986

[8] In contrast to the linear time complexity of type-transparent LAG/DBS (TCS'92).

[9] FoCL 9.4, 10.4, 10.5, specifically 10.5.5.

1.9 Four Kinds of Type-Token Relations

The interaction between the DBS agent's computational cognition and its cognition-external surroundings is based on the pattern-matching of concepts. Recognition is a concept type matching raw data, resulting in a token stored in short term memory. Action is adapting a type to a purpose, resulting in a token realized as raw data.

DBS uses the type-token relation directly for elementary proplets of the semantic kinds concept, but indirectly for indexicals and names, and for complex contents of declarative, interrogative, and imperative sentences.

1.9.1 TYPE AND TOKEN OF A CONCEPT

$$
\begin{array}{ll}
\textit{type} & \textit{token} \\
\begin{bmatrix}
\text{sur: Hund} \\
\text{noun: dog} \\
\text{cat: def sg} \\
\text{sem:} \\
\text{fnc:} \\
\text{mdr:} \\
\text{nc:} \\
\text{pc:} \\
\text{prn:}
\end{bmatrix}
&
\begin{bmatrix}
\text{sur: Hund} \\
\text{noun: dog} \\
\text{cat: def sg} \\
\text{sem:} \\
\text{fnc: snore} \\
\text{mdr:} \\
\text{nc:} \\
\text{pc:} \\
\text{prn: 24}
\end{bmatrix}
\end{array}
$$

The attributes fnc and prn of the type have no value, while those of the token have the values snore and 24. The sur value is from German.

1.9.2 TYPE AND TOKEN OF AN INDEXICAL

$$
\begin{array}{lll}
\textit{type} & \textit{token} & \text{STAR} \\
\begin{bmatrix}
\text{sur: you} \\
\text{noun: pro2} \\
\text{cat: sp2} \\
\text{sem:} \\
\text{fnc:} \\
\text{mdr:} \\
\text{nc:} \\
\text{pc:} \\
\text{prn:}
\end{bmatrix}
&
\begin{bmatrix}
\text{sur: you} \\
\text{noun: pro2} \\
\text{cat: sp2} \\
\text{sem:} \\
\text{fnc: see} \\
\text{mdr:} \\
\text{nc:} \\
\text{pc:} \\
\text{prn: 24}
\end{bmatrix}
\quad \cdots\cdots \quad
\begin{bmatrix}
\text{S: veranda} \\
\text{T: Monday} \\
\text{A: John} \\
\text{R: Mary} \\
\text{prn: 24}
\end{bmatrix}
\end{array}
$$

The type has no prn value and no STAR to point at, while the token has the prn value 24 and may point at the STAR value John or Mary, depending on the syntax.

1.9.3 TYPE AND TOKEN OF A NAME

type

$$
\begin{bmatrix}
\text{sur: Fido} \\
\text{noun:} \\
\text{cat: snp} \\
\text{sem: m sg} \\
\text{fnc:} \\
\text{mdr:} \\
\text{nc:} \\
\text{pc:} \\
\text{prn:}
\end{bmatrix}
$$

token

$$
\begin{bmatrix}
\text{sur: Fido} \\
\text{noun: [dog x]} \\
\text{cat: snp} \\
\text{sem: m sg} \\
\text{fnc:} \\
\text{mdr:} \\
\text{nc:} \\
\text{pc:} \\
\text{prn:24}
\end{bmatrix}
$$

The type has no prn value and the core attribute noun has no 'named referent', while the token has the prn value 24 and the core attribute has the named referent [dog x].

Finally consider the type and token of a DBS proposition,[10] defined as a content. The syntactic mood is specified by the verb's cat value decl as a declarative.

1.9.4 TYPE OF A CONTENT

type

$$
\begin{bmatrix}
\text{sur:} \\
\text{noun: dog} \\
\text{cat: snp} \\
\text{sem: def sg} \\
\text{fnc: find} \\
\text{mdr:} \\
\text{nc:} \\
\text{pc:} \\
\text{prn: K}
\end{bmatrix}
\begin{bmatrix}
\text{sur:} \\
\text{verb: find} \\
\text{cat: \#n}' \text{ \#a}' \text{ decl} \\
\text{sem: past ind} \\
\text{arg: dog bone} \\
\text{mdr:} \\
\text{nc:} \\
\text{pc:} \\
\text{prn: K}
\end{bmatrix}
\begin{bmatrix}
\text{sur:} \\
\text{noun: bone} \\
\text{cat: snp} \\
\text{sem: indef sg} \\
\text{fnc: find} \\
\text{mdr:} \\
\text{nc:} \\
\text{pc:} \\
\text{prn: K}
\end{bmatrix}
$$

This content is a type because there is no STAR and the prn value is a variable, here K. It is a nonlanguage content because the sur slots are empty.

1.9.5 CORRESPONDING TOKEN

token

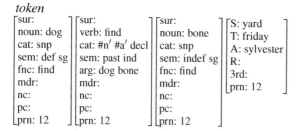

This content is a token because the three content proplets and the STAR proplet are connected by a common prn constant, here 12. According to the STAR, the content resulted as an observation by the agent Sylvester on Friday in the yard.

---

[10] An elementary proposition is a content which uses exactly one prn value.

In summary, the types of individual proplets are lexical word form analyses which are provided by the on-board memory for automatic word form recognition/production. The type of a complex content results from concatenating proplet types with the semantic relations of structure. The content type of a proposition is turned into a content token by adding a STAR and replacing the prn variables with constants (simultaneous substitution).

1.10 Conclusion

In computer science, the input-output distinction holds (i) between a system and its external environment and (ii) between interacting components within a system. The sign-based substitution-driven ontology of Phrase Structure Grammar (PSG) avoids input-output interaction with the system-external reality by using the same S node like a start button as input for the random generation of all the different grammatical structures in the fragment. The agent-based data-driven ontology of DBS, in contrast, provides external nonlanguage input-output in (i) action and (ii) recognition between agents and their environment, and external language input and output between agents in the (iii) speak and (iv) hear modes.

The input to the (iii) speak mode is a hierarchical content and the output a linear surface. The input to the (iv) hear mode is a linear surface and the output a hierarchical content. The challenge for a functionally complete, scientific computational reconstruction of natural language communication is a bidirectional conversion between a linear and a hierarchical coding of the semantic relations of structure.

In DBS, the speak mode turns hierarchical input contents into linear output surfaces by *navigating* along the semantic relations of structure in the input. The hear mode turns linear input surfaces into hierarchical output content by incremental time-linear *syntactic-semantic composition* between the sentence start, defined as a set of proplets already connected (at least partially), and the next word, re-introducing the classical semantic relations of subject/predicate, object\predicate, modifier|modified, and conjunct−conjunct, coded by address.

2. Laboratory Set-Up of Database Semantics

The analysis of natural language in today's linguistics, analytic philosophy, and computer science is either (i) agent-based data-driven or (ii) sign-based substitution-driven. A sign-based ontology has the apparent advantage that it obviates any need for an interface component with sensors for vision and audition, and actuators for manipulation and vocalization. In an age when artificial vision, audition, manipulation, locomotion, and computers did not exist, this was a necessity. The question is how to adjust today's language research to the age of computers and artificial intelligence by changing from a sign-based substitution-driven to an agent-based data-driven ontology?

2.1 Early Times

The absence of computers did not stop the grammarians of the ancient and recent past from contributing essential notions representing important insights, such as accusative, active, adjective, agglutination, agreement, allomorph, analytic, aorist, argument, clause, comparation, conjunction, dative, determiner, domain, ergative, event, function, future, genitive, imperfect, inflection, isolating, medium, modifier, morpheme, morphology, nominative, noun, object, passive, perfect, phrase, pragmatics, predicate, pronoun, proposition, range, relation, semantics, sentential mood, subclause, subject, syntax, synthetic, tense, unaccusative, and verbal mood. Without these notions most of modern linguistics would be unthinkable.

A recent attempt to bring language science into modern times was Chomsky's Generative Grammar (GG) for characterizing the innate universal structure of natural language (nativism). Rewrite rules generate constituent structures from the S node (for sentence or start) by repeated substitution, resulting in phrase structure trees which are defined in terms of the non-semantic notions "dominance" and "precedence," complemented by "government" and "binding." By adding a transformation component to a context-free phrase structure base, the computational complexity of Transformational Grammar increased from polynomial to undecidable (Peters and Ritchie 1973).

Chomsky emphasized repeatedly that GG was "not intended" for modeling communication: "To avoid what has been a continuing misunderstanding, it is perhaps worthwhile to reiterate that a generative grammar is not a model for a speaker or a hearer." (Chomsky 1965, p. 9). Yet, as shown by the analogy with anatomy, it is unlikely that

R. Hausser, *Ontology of Communication*, https://doi.org/10.1007/978-3-031-22739-4_2

a supposedly innate universal model of natural language would be without a speak mode, a hear mode, and a transfer channel.

In consequence, many linguists moved (or returned) from nativism to the study of large data and statistics (Church&Mercer 1993). However, statistics alone is not sufficient for building a talking robot. In analogy, if the Martians came to Earth and modeled cars statistically, the cars would never run. Instead, the Martians would have to chose a car in good running condition, take it apart, study the parts and the functional flow, and reconstruct the mechanisms of the motor, the wheels, the breaks, the transmission, etc., until the reassembled vehicle would run again.

2.2 Study of the Language Signs

A truly classic pioneer of modern linguistics was the Swiss linguist Ferdinand de Saussure (1857–1913), who formulated the most important properties of the natural language signs as the *premier principe*, l'arbitraire de signe,[1] and the *second principe*, caractère linéaire du signifiant.[2] These principles are as valid today as when they were first proposed.

Regarding the second principle, de Saussure continues in good humor:

> Ce principe est évident, mais il semble qu'on ait toujours négligé de l'énoncer, sans doute parce qu'on l'a trouvé trop simple; cependent il est fondamental et les conséquences en sont incalculables; son importance est égale à celle de la première loi. Tout le mécanisme de la langue en dépend.[3]

Ignoring time-linearity is one of those aberrations which are so frequent in the history of science and which often take several centuries to be rectified.

The first attempt at combining time-linearity with detailed grammatical analysis and efficient computation was NEWCAT:

2.2.1 NEWCAT PARSE OF Fido dug the bone up. (CoL 3.3.4)

```
*  (z Fido dug the bone up \.)

   Linear Analysis:

   *START
   1
      (N-H) FIDO
      (N A UP V) DUG
   *NOM+FVERB
   2
      (A UP V) FIDO DUG
      (GQ) THE
   *FVERB+MAIN
   3
      (GQ UP V) FIDO DUG THE
      (S-H) BONE
   *DET+NOUN
   4
      (UP V) FIDO DUG THE BONE
      (UP NP) UP
   *FVERB+MAIN
   5
      (V) FIDO DUG THE BONE UP
      (V DECL) .
   *CMPLT
   6
      (DECL) FIDO DUG THE BONE UP .
```

The grammatical analysis is a formatted *trace* of the computational operations. Each numbered derivation step consists of a sentence start, e.g., (A UP V) FIDO DUG, the next word (GQ) THE, the rule name *FVERB+MAIN, a number (here 2), and the resulting output (GQ UP V) FIDO DUG THE, which redoubles as the input to the next derivation step. As a direct reflection of the computational application of the grammar rules, tracing is the ultimate form of *type transparency* (Berwick and Weinberg 1984). Computational tracing as the exclusive method of grammatical analysis is used in all subsequent work of what became DBS.

Like NEWCAT, 'Computation of Language' (CoL[4]) is still sign-based, but expands the time-linear NEWCAT approach to computational complexity analysis. For example, the formal language $a^k b^k c^k$ is context-sensitive in the PSG hierarchy and parses in exponential time, but is a C1 language in the LAG hierarchy and parses in linear time[5] (CoL 6.4.3, FoCL 10.2.2, TCS'92):

2.2.2 LA GRAMMAR FOR $a^k b^k c^k$

$LX =_{def} \{[a\,(a)], [b\,(b)], [c\,(c)]\}$
$ST_s =_{def} \{[(a)\,\{r_1, r_2\}]\}$
$r_1: (X)\quad (a)\ \Rightarrow (aX)\ \{r_1, r_2\}$
$r_2: (aX)\ (b)\ \Rightarrow (Xb)\ \{r_2, r_3\}$
$r_3: (bX)\ (c)\ \Rightarrow (X)\ \{r_3\}$
$ST_F =_{def} \{[\varepsilon\ rp_3]\}.$

A lexical entry like [a(a)] in the set LX consists of a surface, here a, and a category, here (a). The set ST_s happens to contain only one start state, namely $\{[(a)\,\{r_1, r_2\}]\}$; this means that the first input must have the category (a), i.e., it must have the surface a, and that the rules applying to the first and the second input are limited by the rule package to r_1 and r_2. Rule r_1 adds an (a), r_2 subtracts an (a) and adds a (b), while r_3 subtracts a (b) from the category.

The rule package of r_1 is $\{r_1, r_2\}$, i.e., after r_1 has applied, r_1 and r_2 are tried on the next word, and accordingly for the rules packages of r_2 and r_3. The set ST_F contains

---

[1] 'First principle: arbitrariness of the sign.' It refers to the fact that different languages may use different surfaces, e.g., fauteuil, sessel, and poltrona, for the same kind of thing, here 'easy chair,' based on different *conventions* within the different language communities.

[2] 'Second principle: linear character of the sign.' It refers to the fact that language signs follow each other in a certain grammatical order. Changing the order results in a change of meaning or in ungrammaticality.

[3] 'This principle is obvious, but it seems that stating it explicitly has always been neglected, doubtlessly because it is considered too simple. It is, however, a fundamental principle and its consequences are incalculable. Its importance equals that of the first law. All the mechanisms of language depend on it.' De Saussure ([1916]1972, p. 103)

[4] The software for CoL was written at the Language Technology Institute, Carnegie Mellon University, Pittsburgh in 1986-1988. Thanks to Prof. Carbonell, then director of the LTI, for his generous hospitality and help with Framekit+, and Eric Nyberg, Teruko Mitamura, and Todd Kaufmann for their help in writing the Lisp code.

[5] The term 'time-linear' refers to a grammatical derivation order while the term 'linear time' refers to a computational complexity degree.

only one final state, namely $\{[\varepsilon \; rp_3]\}$, i.e., the category must be empty (ε) and the currently activated rule package must be that of r_3.

Compared to the context-sensitive PSG (FoCL 8.3.7), the LAG is exceedingly plain. Furthermore, the LA Grammars for context-free $a^k b^k$ (CoL 10.2.3) and for context-sensitive $a^k b^k c^k$ are in the same language class of DBS and the number of coefficients, as in $a^k b^k c^k d^k$, $a^k b^k c^k d^k e^k$, etc., has no effect on the linear complexity of their LA Grammars. Like the natural language analysis 2.2.1, an $a^k b^k c^k$ expression is analyzed as a formatted trace of the parse, shown here with the automatic rule counter switched on:

2.2.3 SAMPLE DERIVATION OF **aaabbbccc** WITH ACTIVE RULE COUNTER

```
* (z a a a b b b c c c)
;  1: Applying rules (RULE-1 RULE-2)
;  2: Applying rules (RULE-1 RULE-2)
;  3: Applying rules (RULE-1 RULE-2)
;  4: Applying rules (RULE-2 RULE-3)
;  5: Applying rules (RULE-2 RULE-3)
;  6: Applying rules (RULE-2 RULE-3)
;  7: Applying rules (RULE-3)
;  8: Applying rules (RULE-3)
; Number of rule applications: 14.

   *START-0
   1
      A (A)
      A (A)
   *RULE-1
   2
      A A (A A)
      A (A)
   *RULE-1
   3
      A A A (A A A)
      B (B)
   *RULE-2
   4
      A A A B (A A B)
      B (B)
   *RULE-2
   5
      A A A B B (A B B)
      B (B)
   *RULE-2
   6
      A A A B B B (B B B)
      C (C)
   *RULE-3
   7
      A A A B B B C (B B)
      C (C)
   *RULE-3
   8
```

```
A A A B B B C C (B)
C (C)
*RULE-3
9
A A A B B B C C C (NIL)
```

Expressions which are not in the language, e.g., aaabbc, are analyzed to the point of the ungrammatical continuation, here aaabb+c, and rejected as such. While PSG derivations are substitution-driven by always starting with the same S symbol followed by random applications of rewrite rules (computing possible substitutions), LAG derivations are data-driven by processing the input surfaces one after the other (computing possible continuations). The LAG hierarchy is the first, and so far the only, complexity hierarchy which is orthogonal to the PSG hierarchy.

2.3 Using Successful Communication for the Laboratory Set-Up

In face to face dialogue, the hearer's interpretation begins with the speaker's first word. From there, the hearer follows the sequence of incoming surfaces incrementally, with the speaker at least one word ahead. In indirect communication based on writing or recorded message, there is no limit on the speaker's lead.

This could be taken as a reason for starting the scientific analysis of natural language communication with the speak mode. However, there is a more important aspect to the distinction between the two modes, namely the difference in the respective input and output: the speak mode takes a cognitive content as input and produces an external surface as output, while the hear mode takes an external surface as input and produces a cognitive content as output.[6]

For a scientific analysis and reconstruction of language communication, the hear mode has the advantage of concrete external input, i.e., the raw data of the language-dependent surfaces. They have no meaning or grammatical properties (no reification in DBS), but they are measurable by natural science and interpretable by automatic speech recognition (asr) or optical character recognition (ocr). The input to the speak mode, in contrast, is agent-internal cognitive content which can only be inferred.

Therefore DBS starts the computational reconstruction of natural language communication with the hear mode's first step, namely automatic word form recognition of the raw surface input by means of computational pattern matching. The output of the hear mode is an agent-internal, purely cognitive structure: it derives the literal meaning$_1$ (PoP-1, FoCL 4.3.3) of the input surface as an agent-internal content.

In order for communication to be successful, the following condition must be fulfilled:

---

[6] The speak mode and the hear mode each require their own software because their input and output are different. In other words, it is impossible to use the same software for runing "upward" in the speak mode (from content to surface) and "downward" for the hear mode (from surface to content). It is different in inferencing, which allows *inductive* (forward) and *abductive* (backward) use by one and the same algorithm because both directions use the same kind of input and output.

2.3.1 MINIMAL CONDITION FOR SUCCESSFUL COMMUNICATION

The meaning$_1$ content used as input by the speak mode and the meaning$_1$ content derived as output in the hear mode must be the same.

This condition is the pivot of the DBS *laboratory set-up*:

2.3.2 DEFINITION OF THE DBS LABORATORY SET-UP

- The content automatically derived as output in the hear mode is reused systematically as the input to the automatic speak mode derivation.[7]
- The content of a given example surface is correct if, and only if, the hear mode's input surface equals the speak mode's output surface.

The laboratory set-up provides a fully automatic, clear and simple method of verification. It requires that (i) the grammatical details of the speak mode suffice for the associated hear mode to automatically derive the speaker's content and (ii) that the grammatical details of hear mode content suffice for the associated speak mode derivation to automatically produce the hearer's surface.

2.3.3 LABORATORY SET-UP: FROM HEAR MODE TO SPEAK MODE

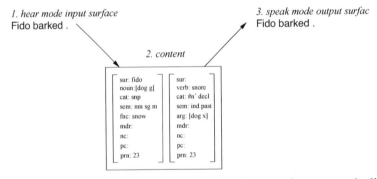

The DBS laboratory set-up uses the hear mode (1, 2) to produce semantically well-motivated output content as input for the speak mode (2, 3). By treating the speak and the hear mode as separate derivations from content to surface and surface to content, the two modes benefit each other.

The laboratory set-up is based on switching off inferencing, temporarily limiting the think-speak mode to traversing meaning$_1$ content and producing literal surface representations in the natural language of choice ('narrative speak mode'). This resembles a sign-based approach in that it excludes the pragmatic aspects of interpretation, but differs in that it has an explicit notion of content and includes the computational reconstruction of the speak and the hear mode. When the direction from speaker to

---

[7] Successful communication requires input-output compatibility in the speak mode.

hearer outside the laboratory set-up is re-established and inferencing for non-literal use is switched back on, the speak mode (deductive) may realize inference content as language-dependent surfaces and the hear mode (abductive) may interpret the surfaces as inference content – data-driven, without any need for additional software.

2.4 From Operational Implementation to Declarative Specification

Following general practice at the time, NEWCAT and CoL use *holistic* loading, i.e., they take a complete expression as input and process it word form by word form in left-associative[8] order. This is suitable for parsing a set of isolated linguistic examples, but not for parsing a text, e.g., a complete novel by Tolstoy.

The processing order of the DBS hear mode also uses the left-associative derivation order, but the loading is *incremental*, i.e., (i) successful application of an operation triggers next word lookup, (ii) which triggers activation of all potential next operations by matching their second input pattern, (iii) which look for a proplet at the now front matching their first input pattern, and (iv) apply if they find one.

Separate treatments of the speak and the hear mode came with a change from the ordered triple analysis of a word form in NEWCAT to the proplet format as a non-recursive feature structure with ordered attributes. For example, the ordered triple analysis of dug in 2.2.1 was changed into the following proplet:

2.4.1 TRANSITION FROM ORDERED TRIPLE TO LEXICAL PROPLET

ordered triple format   *proplet format of DBS*

[dug (N A up V) dig]

$$
\begin{bmatrix}
\text{sur: dug} \\
\text{verb: dig} \\
\text{cat: N' A' up' V} \\
\text{sem: } up \text{ ind past} \\
\text{arg:} \\
\text{mdr:} \\
\text{nc:} \\
\text{pc:} \\
\text{prn:}
\end{bmatrix}
$$

The two formats differ as follows.

2.4.2 COMPARING THE NEWCAT-CoL APPROACH WITH DBS

1. The ordered triple format does not distinguish between valency slots and valency fillers, whereas DBS proplets mark valency slots with ', e.g., N'.
2. In the ordered triple format, valency positions are canceled by deletion (as in Categorial Grammar), whereas the DBS hear mode cancels valency positions by #-marking, thus preserving the information for the speak mode.

---

[8] Aho and Ullman 1977, p. 47.

3. Derivations in the ordered triple format prefer ending on empty category and use the complete derivation as the resulting content. A hear mode derivation in the proplet format, in contrast, results in a content which stands for itself, defined as a set of proplets connected by address, leaving the derivation behind.
4. The proplet format enables string-search-based storage in and retrieval from a content-addressable database contained in a cognitive agent with an interface component (agent-based ontology).
5. The on-board database interacts with the agent's interface component for the recognition and production of language surfaces, as well as the recognition of and action with nonlanguage contents.
6. The agent's on-board orientation system provides the STAR for the interpretation of the sign kind 'indexical.'

The NEWCAT derivation 2.2.1 has the following hear mode derivation in DBS:

2.4.3 DBS HEAR MODE DERIVATION OF Fido dug the bone up.

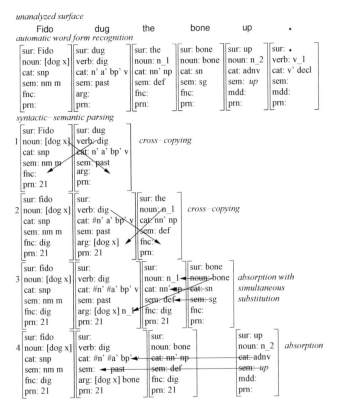

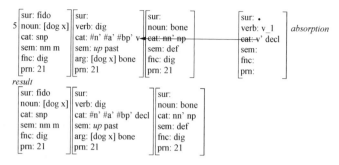

The bare preposition up is absorbed in line 4. It #-cancels the valency position bp' in the third cat slot of *dig* and writes its sem value up into the initial sem slot of the verb, making it available for the speak mode. Like the NEWCAT-style derivation 2.2.1, the derivation exactly mirrors the sequence of operation applications[9] (tracing).

The transition from NEWCAT and CoL to FoCL, NLC, CLaTR, TExer, and CC may be summarized as follows. Instead of taking the whole surface of a sentence or text as input (holistic loading), there is incremental next word lookup.[10] Instead of deleting valency positions in the verb's cat slot, they are preserved for the speak mode by #-canceling. Instead of using the derivation, e.g., 2.2.1, as the grammatical analysis, there results a *content* (e.g., result in 2.4.3) which is not dependent on the hear mode derivation (essential for nonlanguage cognition). Instead of longer and longer intermediate states, there are larger and larger sets of proplets connected by address (order-free), which is essential for the storage and retrieval in the on-board content-addressable database.

2.5 Formal Fragments of Natural Language

There are formal language analyses in the tradition of symbolic logic (Montague 1974) and computational complexity (FoCL) theory which use explicit rule systems to analyze-generate limited 'fragments' of natural or formal languages like $a^k b^k$, $a^k b^k c^k$, $a^k b^k c^k d^k$, etc. (2.2.3). A fragment is precisely defined as a set of examples for the analysis of specific, natural or artificial, grammatical structures. The language data in a fragment are limited, but their analysis is required to be explicit.

The use of software in the computational analysis of fragments opens a new dichotomy as compared to precomputational linguistics, namely between (i) the *declarative specification* and (ii) the *operational implementation*. The declarative specification represents the necessary properties of the software and must be simultaneously suitable (a) for reading by humans and (b) for a straightforward translation into a general purpose programming language of choice. The operational implementation, in contrast, has additional accidental properties, namely those which distinguish equivalent implementations in different programming languages.

---

[9] For the sequence of explicit hear and speak mode operation applications see TExer 4.3.

[10] For the time-linear transition from one sentence to the next see TExer 2.1.

After working on implementing a fragment of natural language, there naturally arises a scientific interest in leaving the accidental properties behind and work out the necessary ones in the systematic format of a declarative specification.[11] Conversely, after working on a declarative specification for a fragment of a natural language, there naturally arises a scientific interest in verifying the fragment in the form of an operational implementation.[12]

2.6 Incremental Upscaling Cycles

Once a current fragment has been supplied with a declarative specification for the speak and the hear mode, and been verified by an operational implementation, the next upscaling cycle is started by extending the current fragment with a limited number of additional examples which have new and interesting syntactic and semantic properties. For this kind of work, a standard computer of today is sufficient. It provides the keyboard for input and the screen for output, which allows to implement the hear mode, the content-addressable database with its now front mechanism, and the think-speak mode navigation with and without surface realization, using placeholders for concepts.[13]

2.7 Conclusion

For building a talking robot, the recognition and action hardware of the interface component should be co-developed with the cognition software. This holds specifically for building the on-board orientation system and for supplying the concept placeholders with procedural implementations (CC Chapter 11).

---

[11] The first operational implementations were published as NEWCAT and CoL. Work on the declarative specification began with FoCL, continued with NLC and CLaTR, and culminated for now in TExer and CC.

[12] This kind of work is of practical use, and may serve as an inexhaustible source of computationally demanding and linguistically interesting thesis topics at all degree levels.

[13] The development of the DBS software in Lisp, Java, and C up to the year 2011 is summarized by Handl (2012) in chapters 6 and 7. For a list of the natural languages analyzed in DBS and the researchers who did the work see CLaTR 3.5.3.

3. Outline of DBS

DBS models the cycle of natural language communication as a transition from the hear to the think to the speak and back to the hear mode (turn taking). In contradistinction to the sign-based substitution-driven approaches of truth-conditional semantics and phrase structure grammar, DBS is agent-based and data-driven. The goal is an efficient computational theory of natural language communication suitable for a talking autonomous robot.

Instead of denoting truth values, propositions *are content* in DBS. Content is built from the classical semantic kinds, i.e., referent, property, and relation, which are concatenated by the classical semantic relations of structure, i.e., functor-argument and coordination. To enable reference as an agent-internal cognitive process, language and nonlanguage contents use the same computational data structure and operation kinds, and differ mostly in the presence vs. absence of language-dependent surfaces.

DBS consists of (i) an interface component which takes raw data as input (recognition) and produces raw data as output (action); (ii) an on-board database for storing and retrieving content provided by recognition, inferencing, and action; (iii) a now front as the arena for processing current content; (iv) an on-board orientation system (OBOS); and (v) an operations component for (a) content activation and inferencing in the think mode, (b) surface-content mapping in the hear mode, and (c) content-surface mapping in the speak mode.

3.1 Building Content in the Agent's Hear Mode

DBS defines a content in terms of concepts like square (3.7.1) or blue (3.7.3) connected with the classical semantic relations of structure, i.e. subject/predicate, object\predicate, modifier|modified, and conjunct−conjunct. The concepts are supplied by the agent's memory and defined as types. In recognition, they are activated by matching raw data provided by the interface component, resulting in tokens.[1] In action, a type is adapted to a purpose as a token and realized as raw data (3.7.2, 3.7.4).

---

[1] The type-token terminology was introduced by C. S. Peirce (CP 4:537). It goes back to Aristotle's distinction between the *necessary* and the *accidental*.

© The Author(s), under exclusive license to Springer Nature Switzerland AG 2023
R. Hausser, *Ontology of Communication*, https://doi.org/10.1007/978-3-031-22739-4_3

For concatenation, concepts are embedded as core values into nonrecursive feature structures with ordered attributes, called proplets. The semantic relations between proplets are established by address, making proplets order-free for purposes of storage and retrieval in the agent's content-addressable on-board database. Proplets serve as the computational data structure of DBS.

3.1.1 THE CONTENT OF The dog snored.

$$
\begin{bmatrix}
\text{sur:} \\
\text{noun: } \textbf{dog} \\
\text{cat: def sg} \\
\text{sem:} \\
\text{fnc: } \textbf{snore} \\
\text{mdr:} \\
\text{nc:} \\
\text{pc:} \\
\text{prn: 24}
\end{bmatrix}
\begin{bmatrix}
\text{sur:} \\
\text{verb: } \textbf{snore} \\
\text{cat: \#n' decl} \\
\text{sem: past ind} \\
\text{arg: } \textbf{dog} \\
\text{mdr:} \\
\text{nc:} \\
\text{pc:} \\
\text{prn:24}
\end{bmatrix}
$$

The proplets implement the subject/predicate relation by using the noun value **dog** of the first proplet as the arg value of the second, and the verb value **snore** as the fnc value of the first (bidirectional pointering). The order-free proplets of a content are stored and retrieved according to the alphabetical sequence of their core values, yet are connected by the address of their continuation values, here (snore 24) and (dog 24). In the hear mode, the content 3.1.1 results from the following derivation:

3.1.2 SURFACE-COMPOSITIONAL TIME-LINEAR DERIVATION

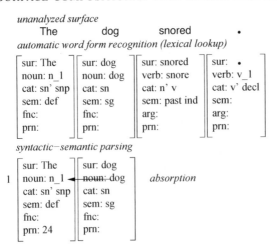

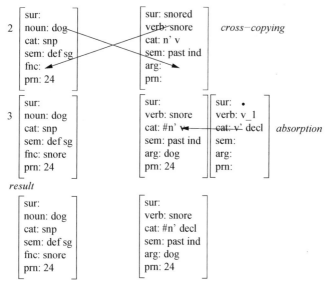

The operations for concatenation in the hear mode, and activation and inferencing in the think-speak mode consist of (i) an antecedent, (ii) a connective, and (iii) a consequent. Defined as proplet patterns, operations are data-driven in that they are activated by matching content proplets.[2].

The hear mode uses three kinds of operations, each characterized by a connective: (1) × for cross-copying, (2) ∪ for absorption, and (3) ∼ for suspension. Cross-copying encodes the semantic relations such as SBJ×PRED (line 2). Absorption combines a function word with a content word such as DET∪CN (line 1) or another function word as in PREP∪DET. Suspension such as ADV∼NOM (TExer 3.1.3) applies if no semantic relation exists for connecting the next word with the content processed so far, as in Perhaps ∼ Fido (slept).

Consider the hear mode operation SBJ×PRED as it applies in line 2 of 3.1.2:

3.1.3 HEAR MODE APPLYING SBJ×PRED APPLYING (CROSS-COPYING)

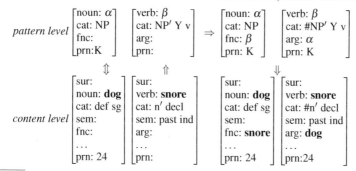

---

[2] While the hear mode takes word form surfaces as input, the input to the think mode is content. Think mode operations are used with and without a surface realization, depending on whether the language-dependent lexicalization rules in the sur slots of the output pattern are turned on or off (3.3.2).

The second input proplet to a hear mode operation is the 'next word' provided by automatic word form recognition, here *snore*. By matching the next word proplet and the second input pattern at the pattern level (⇑), the operation is triggered to look for a content proplet matching its first input pattern (⇕) at the now front (3.2.3). By binding α to dog and β to snore, the consequent produces the output as content proplets (⇓).

3.2 Storage and Retrieval of Content in the On-Board Memory

Contents derived in the hear mode and activated in the think-speak mode (3.3) have in common that they are defined as sets of self-contained proplets, concatenated by proplet-internal address. As sets, the proplets of a content are order-free, which is essential for their storage in and retrieval from the agent's A-memory (formerly called word bank). The database schema of A-memory is defined as follows:

3.2.1 TWO-DIMENSIONAL DATABASE SCHEMA OF A-MEMORY

- *horizontal token line*
  Horizontally, proplets with the same core value are stored in the same token line in the time-linear order of their arrival.
- *vertical column of token lines*
  Vertically, token lines are in the alphabetical order induced by the letter sequence of their core value.

In the hear mode, the arrival order of proplets is recorded by (a) the position in their token line and by (b) their prn value. The *(i) member proplets* are followed by a free slot as part of the column called the *(ii) now front*, and by the *(iii) owner*:[3]

3.2.2 A-MEMORY BEFORE INCREMENTAL STORAGE OF 3.1.1

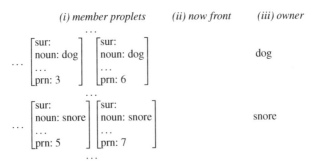

The owners equal the core values in their token line and are used for access in storage and retrieval. Proplets provided by current recognition, by A-memory, or by inferencing are stored at the now front in the token line corresponding to their core value:

3.2.3 STORAGE OF 3.1.1 AT THE NOW FRONT OF A-MEMORY

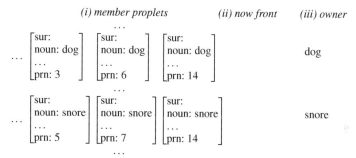

(i) member proplets   *(ii) now front*   *(iii) owner*

$$
\ldots
\begin{bmatrix} \text{sur:} \\ \text{noun: dog} \\ \ldots \\ \text{prn: 3} \end{bmatrix}
\begin{bmatrix} \text{sur:} \\ \text{noun: dog} \\ \ldots \\ \text{prn: 6} \end{bmatrix}
\quad
\begin{bmatrix} \text{sur: chien} \\ \text{noun: dog} \\ \ldots \\ \text{prn: 14} \end{bmatrix}
\qquad \text{dog}
$$

$$
\ldots
\begin{bmatrix} \text{sur:} \\ \text{noun: snore} \\ \ldots \\ \text{prn: 5} \end{bmatrix}
\begin{bmatrix} \text{sur:} \\ \text{noun: snore} \\ \ldots \\ \text{prn: 7} \end{bmatrix}
\quad
\begin{bmatrix} \text{sur: ronfler} \\ \text{noun: snore} \\ \ldots \\ \text{prn: 14} \end{bmatrix}
\qquad \text{snore}
$$

Once a content has been assembled as a proposition, the now front is cleared by moving it and the owners to the right into fresh memory space (loom-like clearance, 3.3). This leaves the proplets of the current content behind in what is becoming their permanent storage location as member proplets never to be changed, like sediment.

3.2.4 A-MEMORY AFTER NOW FRONT CLEARANCE

(i) member proplets   *(ii) now front*   *(iii) owner*

$$
\ldots
\begin{bmatrix} \text{sur:} \\ \text{noun: dog} \\ \ldots \\ \text{prn: 3} \end{bmatrix}
\begin{bmatrix} \text{sur:} \\ \text{noun: dog} \\ \ldots \\ \text{prn: 6} \end{bmatrix}
\begin{bmatrix} \text{sur:} \\ \text{noun: dog} \\ \ldots \\ \text{prn: 14} \end{bmatrix}
\qquad \text{dog}
$$

$$
\ldots
\begin{bmatrix} \text{sur:} \\ \text{noun: snore} \\ \ldots \\ \text{prn: 5} \end{bmatrix}
\begin{bmatrix} \text{sur:} \\ \text{noun: snore} \\ \ldots \\ \text{prn: 7} \end{bmatrix}
\begin{bmatrix} \text{sur:} \\ \text{noun: snore} \\ \ldots \\ \text{prn: 14} \end{bmatrix}
\qquad \text{snore}
$$

Current now front clearance is triggered when its proplets have ceased to be candidates for additional processing, i.e., when an elementary proposition is completed (formally indicated by the automatic incrementation of the prn value for the next proposition). Exceptions arise in *extrapropositional* (i) coordination and (ii) functor-argument. In these two cases, the verb of the completed proposition remains at the now front for cross-copying with the verb of the next proposition until the extrapropositional relation has been established in the strictly time-linear derivation order of DBS.

3.3 Speak Mode Riding Piggyback on the Think Mode

The speak mode counterpart to the hear mode derivation 3.1.2 is a graphical characterization of the semantic relations of structure, here for subject/predicate:

---

[3] The terminology of member proplets and owner values is reminiscent of the member and owner records in a classic network database (Elmasri and Navathe [1989] 2017), which inspired the content-addressable database schema of the A-memory in DBS.

3.3.1 SEMANTIC RELATIONS GRAPH UNDERLYING THE CONTENT 3.1.1

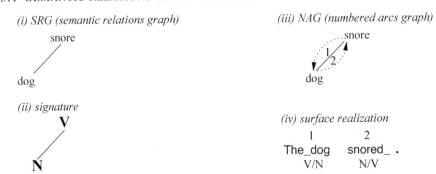

(i) SRG (semantic relations graph)

snore

dog

(ii) signature

V

N

(iii) NAG (numbered arcs graph)

snore

dog

(iv) surface realization

1	2
The_dog	snored_ .
V/N	N/V

The static aspects of the semantic relations of structure are shown on the left: the *(i) SRG* is based on the core values of the content and the *(ii) signature* on the core attributes. The dynamic aspects of a think-speak mode activation are shown on the right: the arc numbers of the *(iii) NAG* are used for specifying a time-linear think mode navigation along the semantic relations between proplets. The *(iv) surface realization* shows the language-dependent production as the speak mode riding piggy-back on the think mode navigation.

The think mode uses the following kinds of traversal operations: (1) predicate/subject, (2) subject/predicate, (3) predicate\object, (4) object\predicate, (5) noun↓adnominal, (6) adnominal↑noun, (7) verb↓adverbial, (8) adverbial↑verb, (9) noun→noun, (10) noun←noun, (11) verb→verb, (12) verb←verb, (13) adnominal→adnominal, and (14) adnominal←adnominal.

The think mode operations driving the traversal of the NAG in 3.3.1 are V/N and N/V, and apply as follows (shown with English surface production):

3.3.2 NAVIGATING WITH V/N FROM *snore* TO *dog* (arc 1)

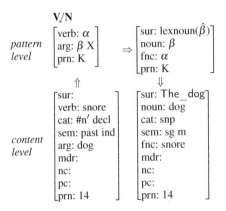

V/N

pattern level

$$\begin{bmatrix} \text{verb: } \alpha \\ \text{arg: } \beta\ X \\ \text{prn: K} \end{bmatrix} \Rightarrow \begin{bmatrix} \text{sur: lexnoun}(\hat{\beta}) \\ \text{noun: } \beta \\ \text{fnc: } \alpha \\ \text{prn: K} \end{bmatrix}$$

content level

$$\begin{bmatrix} \text{sur:} \\ \text{verb: snore} \\ \text{cat: \#n' decl} \\ \text{sem: past ind} \\ \text{arg: dog} \\ \text{mdr:} \\ \text{nc:} \\ \text{pc:} \\ \text{prn: 14} \end{bmatrix} \begin{bmatrix} \text{sur: The\_dog} \\ \text{noun: dog} \\ \text{cat: snp} \\ \text{sem: sg m} \\ \text{fnc: snore} \\ \text{mdr:} \\ \text{nc:} \\ \text{pc:} \\ \text{prn: 14} \end{bmatrix}$$

3.3.3 Navigating with N/V from *dog* back to *snore* (arc 2)

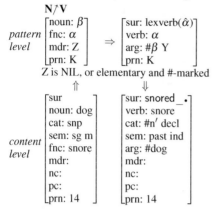

$$\begin{array}{c} \mathbf{N/V} \\ \textit{pattern level} \begin{bmatrix} \text{noun: } \beta \\ \text{fnc: } \alpha \\ \text{mdr: Z} \\ \text{prn: K} \end{bmatrix} \Rightarrow \begin{bmatrix} \text{sur: lexverb}(\hat{\alpha}) \\ \text{verb: } \alpha \\ \text{arg: } \#\beta \text{ Y} \\ \text{prn: K} \end{bmatrix} \end{array}$$

Z is NIL, or elementary and #-marked

$$\textit{content level} \begin{bmatrix} \text{sur} \\ \text{noun: dog} \\ \text{cat: snp} \\ \text{sem: sg m} \\ \text{fnc: snore} \\ \text{mdr:} \\ \text{nc:} \\ \text{pc:} \\ \text{prn: 14} \end{bmatrix} \quad \begin{bmatrix} \text{sur: snored}\_. \\ \text{verb: snore} \\ \text{cat: } \#n' \text{ decl} \\ \text{sem: past ind} \\ \text{arg: } \#\text{dog} \\ \text{mdr:} \\ \text{nc:} \\ \text{pc:} \\ \text{prn: 14} \end{bmatrix}$$

If the lexnoun rules in the sur slot of the output patterns are switched on (as assumed in the surface realization of 3.3.1), they generate a language-dependent surface using relevant values of the output proplet.

3.4 Component Structure of Cognition

The component structure of DBS cognition may be summarized as follows:

3.4.1 Two-dimensional layout of DBS cognition components

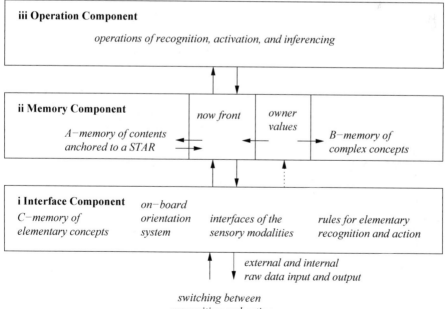

Cognitive content is processed at the now front. It gets proplets (a) from the interface component (aided by the owners) and (b) from A-memory. For processing, the now front provides proplets as input to (iii) the operations, which either replace the input with their output or add their output to the input. As the now front is cleared in regular intervals by moving into fresh memory space (3.2), the processed proplets are left behind in A-memory like sediment. Processing may also result in blueprints for action, which may be copied to the interface component for realization as raw data (subjunctive transfer, CLaTR 5.6).

3.5 Sensory Media, Processing Media, and Their Modalities

A talking autonomous robot and its human prototype use different processing media, mockingly called hardware vs. wetware. Consequently, adequate modeling is limited to *functional equivalence*. A classic example of independence from the processing medium is the basic operations of arithmetic: 3+4 equals 7 no matter whether the calculation is performed by (i) a human,[4] (ii) a mechanical calculator, or (iii) a computer.

In addition to the processing media there are the sensory media. In natural language communication, there exist four, each of which has two sensory modalities.[5] For example, if the speaker chooses the medium of speech, the only sensory modality for production is vocalization ($\searrow$), which leaves the hearer no other option than using the sensory modality of audition ($\nearrow$). This asymmetry of modalities holds also for the other sensory media of natural language, namely writing, Braille, and sign language:

3.5.1 SENSORY MEDIA AND THEIR MODALITIES IN COMMUNICATION

In terms of human evolution, the primary sensory medium is speech.

While the sensory media must be the same in the natural prototype and the artificial counterpart, as required by functional equivalence, the processing media are fundamentally different between the two. For the natural prototype, neurology suggests an electrochemical processing medium, though much is still unknown.[6] In artificial DBS cognition, in contrast, the processing medium is a programming language; its processing modalities are (i) the declarative specification of commands for interpretation by the computer and (ii) their procedural execution by the computer's electronic operations.

---

[4] The operations of arithmetic as they are processed by the human brain are described by Menon (2011).

[5] In the literature, the term modality has a multitude of uses, such as the temperature (Dodt and Zotterman 1952), the logical (Barcan Marcus 1961), and the epistemic (Kiefer 2018) modalities.

[6] For an early overview see Benson (1994).

3.5.2 Processing Media and their Processing Modalities

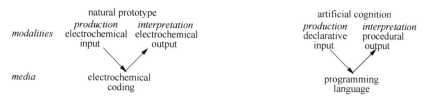

Utilizing a programming language as the processing medium of an artificial agent requires an interface component capable of efficiently mediating between raw data and an alphanumeric representation in recognition and action.

3.6 Reference as a Purely Cognitive Process

Sign-based philosophy defines reference as a relation between language (referring part) and the world (referred-to part). Reimer and Michaelson (2014) extend the referring part from language to "representational tokens," which include cave paintings, pantomime, photographs, videos, etc. DBS continues in this direction by generalizing the referring part to content *per se*, i.e., without the need for any cognition-external counterpart (3.6.3, [-surface, -external]).

At the same time, agent-based DBS confines reference to nouns (CC 1.5.3, 12.3.3) and distinguishes (1) between referring nouns with and without external surfaces and (2) between referred-to nouns with and without external[7] counterparts. The two distinctions are characterized by the binary features [±surface] and [±external], whereby [+external] reference is called *immediate*, while [−external] reference is called *mediated* (FoCL 4.3.1).

For example, identifying "the man with the brown coat" (Quine 1960) with someone seen before, or identifying an unusual building with an earlier language content, e.g., something read in a guide book or heard about, is [−surface +external]. Talking about Aristotle or J.S. Bach, in contrast, is [+surface −external].

The [±surface] and [±external] distinctions are not available in truth-conditional semantics and generative grammar because their sign-based ontology does not distinguish (i) between cognition-external reality and cognition-internal processing, and between (ii) recognition and action, including the hear- and the speak-mode distinction. Also, there is no onboard memory (content-addressable database) with an onboard orientation system and no algorithm for moment-by-moment monitoring. In short, sign-based substition-driven systems exclude by definition the components of a von Neumann machine (vNm, von Neumann 1945) and are therefore unsuitable in principle for designing and building a talking robot in particular and AI in general.

Let us go systematically through the four kinds of generalized DBS reference, beginning with the [+surface +external] constellation between speaker and hearer:

---

[7] Newell and Simon (1972) call the agent's external surroundings the *task environment*.

3.6.1 IMMEDIATE REFERENCE IN LANGUAGE COMMUNICATION

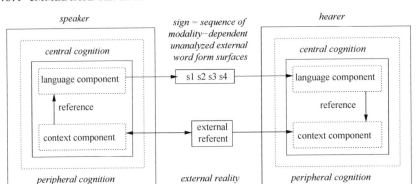

Agent-externally, language surfaces (shown here as $\boxed{\text{s1 s2 s3 s4}}$) are modality-specific unanalyzed external signs (raw data) which are passed from the speaker to the hearer and have neither meaning nor any grammatical properties whatsoever at all (no reification in DBS), but may be measured by the natural sciences.

The corresponding [+surface −external] constellation between the speaker and the hearer is as follows:

3.6.2 MEDIATED REFERENCE IN LANGUAGE COMMUNICATION

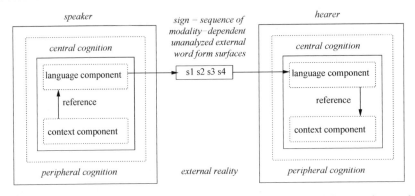

The reference relation begins with content in the memory of the speaker and ends as content in the memory of the hearer. The mechanisms of assigning surfaces to content in the speak mode and content to surfaces in the hear mode are the same in immediate and mediated language reference.

The graphs 3.6.1 and 3.6.2 show the speaker on the left, the sign in left-to-right writing order in the middle, and the hearer on the right. This is a possible constellation which is in concord with the naive assumption that time passes with the sun from left to right (→) on the Northern Hemisphere. Yet it appears that the first surface s1 leaves the speaker last and the last surface s4 arrives at the hearer first, which would be functionally incorrect.

It is a pseudo-problem, however, which vanishes if each surface is transmitted individually and placed to the right of its predecessor, i.e., (((s1 s2) s3) s4). This *left-associative*[8] departure and arrival structure allows incremental surface by surface processing, provided the derivation order is based on computing possible continuations, as in Left-Associative Grammar (TCS'92).

Nonlanguage reference differs from language reference in that it is [−surface]. Thereby nonlanguage immediate reference is [−surface +external] while nonlanguage mediated reference is [−surface −external]:[9]

3.6.3 NONLANGUAGE IMMEDIATE VS. MEDIATED REFERENCE

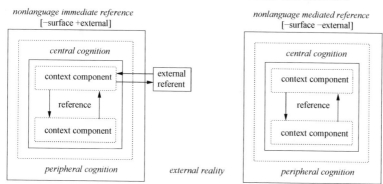

The referring content in the [−surface +external] constellation is a current nonlanguage recognition, as when recognizing a person on the street. In the [−surface −external] constellation of nonlanguage mediated reference, in contrast, the referring content is activated without an external trigger, for example, by reasoning. In both, the referred-to content is resonating (CC Sects. 3.2, 3.3) in memory.

Computationally, the conceptual view of reference as a vertical interaction between two separate components in 3.6.1–3.6.3 is implemented as a horizontal relation between two proplets in the same token line:

3.6.4 COMPARING THE NAIVE AND THE COMPUTATIONAL SOLUTION

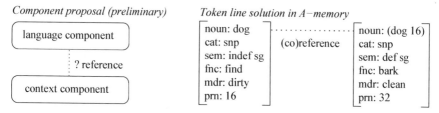

Because the semantic kind of referent is limited to the syntactic kind of noun, (co)reference is restricted to nominal concepts, indexicals, and names (CC 6.4.1,

---

[8] Aho and Ullman (1977, p. 47). Thanks to Profs. Ron Kaplan and Stuart Shieber for pointing it out.

[9] Binary features like [± voiced] are used in the "feature bundles" of Chomsky & Halle (1968).

6.4.4–6.4.6). The core value of the referring noun (shadow, copy) at the now front is always an address. The core value of the referred-to noun (referent, original) is never an address. The fnc and mdr values are free (identity in change, CC 6.4.7).

3.7 Grounding

The semantics of DBS is grounded (Barsalou et al. 2003, Steels 2008, Spranger et al. 2010). In recognition, concept types (supplied by the agent's memory) are matched with raw data (provided by sensors of the agent's interface component):

3.7.1 RECOGNITION OF square

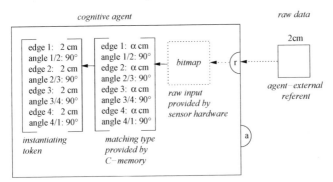

The raw data are supplied by a sensor, here for vision, as input to the interface component. The raw data are matched by the type, resulting in a token.

In action, a type is adapted to a token for the purpose at hand and realized by the agent's actuators as raw data:

3.7.2 ACTION OF REALIZING square

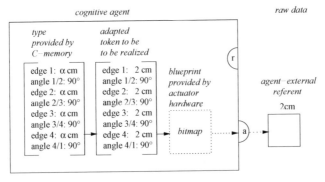

The token is used as a blueprint for action, (e.g., drawing a square).

Next consider the recognition of a color, here blue:

3.7.3 RECOGNITION OF blue

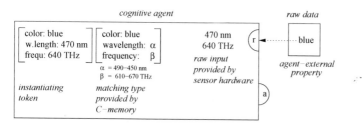

An example of the corresponding action is turning on the color blue, as in a cuttlefish (metasepia pfefferi) using its chromatophores:

3.7.4 ACTION OF REALIZING blue

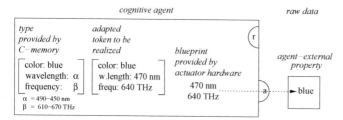

The concept type matches different shades of blue, whereby the variables α and β are instantiated as constants in the resulting token. Recognition and production of the color blue is a general mechanism which may be applied to all colors. It may be expanded to infrared and ultraviolet, and to varying intensity.[10]

3.8 Conclusion

Pattern matching based on the type-token relation applies to nonlanguage items (e.g., 3.7.1, 3.7.2, 3.7.3, 3.7.4) and language surfaces alike. For example, in the surfaces of spoken language the type generalizes over different pitch, timbre, dialect, and speaker-dependent pronunciation. In written language, the type generalizes over the size, color, and font of the letters. Computational type-token matching is more adequate descriptively than the nonbivalent (Rescher 1969; FoCL 20.5) and fuzzy (Zadeh 1965) logics for treating vagueness because type-token matching treats the phenomenon at the root (best candidate principle in pattern matching, FoCL 5.2) instead of tinkering with the truth tables of Propositional Calculus.

---

[10] Complementary approaches from cognitive psychology are prototype theory (Rosch 1975) and Recognition by Components (RBC) based on geons (Biederman 1987).

4. Software Mechanisms of the Content Kinds

The semantics of agent-based DBS is 'grounded' in that the Content kinds *concept, indexical,* and *name* have their foundation in the agent's recognition and action. Each Content kind has its own computational Mechanism. For a concept it is computational pattern *matching* between the type provided by memory and raw data provided by the interface component. For an indexical it is *pointing* at a STAR value of the on-board orientation system (OBOS). For a name it is the address of a 'named referent,' which an implicit or explicit act of *baptism* inserts into a lexical name proplet as the core value.

Orthogonal to the Content kinds and their computational Mechanisms are the (a) Semantic kinds *referent, property,* and *relation* with their associated Syntactic kinds (a) *noun,* (b) *adj* and *intransitive verb,* and (c) *transitive verb.* It is shown that the Semantic kind of referent is restricted to the Syntactic kind of noun, but utilizes the computational Mechanisms of matching, pointing, and baptism. Conversely, figurative use is restricted to the computational Mechanism of matching, but uses the Semantic kinds referent, property, and relation.

4.1 Apparent Terminological Redundancy

The notions noun, verb, and adjective from linguistics (philology) have counterparts in analytic philosophy, namely referent, relation, and property, and in symbolic logic, namely argument, functor, and modifier:

4.1.1 THREE TIMES THREE RELATED NOTIONS

	(a) *linguistics*	(b) *philosophy*	(c) *symbolic logic*
1.	noun	referent (object)	argument
2.	verb	relation	functor
3.	adj	property	modifier

We take it that these variants are not merely different terms for the same things, but different terms for different aspects of the same things. In particular, the linguistic terminology may be viewed as representing the syntactic aspect, the philosophical

R. Hausser, *Ontology of Communication*, https://doi.org/10.1007/978-3-031-22739-4_4

terminology as representing the associated semantic aspect, and the logical terminology as a step towards a computational implementation.

In DBS, the distinctions are related as follows:

4.1.2 1ST CORRELATION: SEMANTIC AND SYNTACTIC KIND

Semantic kind	*Syntactic kind*
1. referent	noun
2. property	adn, adv, adnv, intransitive verb
3. relation	transitive verb

The Semantic kinds *referent, property,* and *relation* correspond to argument, 1-place functor, and 2- or 3-place functor, respectively, in Symbolic Logic. Syntactically, *property* splits up into adn, adv, adnv (including prepnouns), and 1-place verb. *Relation* splits up into 2- and 3-place verbs.

The distinction between (i) Semantic and (iii) Syntactic kinds is complemented by a second, orthogonal pair of triple distinctions, namely (ii) Content kinds and associated (iv) computational Mechanisms:

4.1.3 2ND CORREL.: CONTENT KINDS AND COMPUT. MECHANISMS.

Content kind	*computational Mechanism*
a. concept	matching
b. indexical	pointing
c. name	baptism

The terms of the three Content kinds and the correlated Mechanisms have had informal use in the literature,[1] but without an agent-based ontology. The essential points of the Mechanisms in DBS are their obvious computational realizations (4.4–4.6), which have not been utilized until now.

The dichotomies 4.1.2 and 4.1.3 provide 12 ($2\times2\times3$) basic notions. Empirically, they combine into six classes of proplets which constitute the semantic building blocks of DBS cognition in general and natural language communication in particular. The six classes form what we call the *cognitive square*:

4.1.4 COGNITIVE SQUARE OF DBS

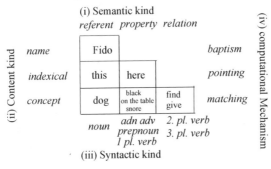

The twelve basic notions of this NLC 2.6.9 extension are distributed over six basic proplets kinds such that no two are characterized the same:

4.1.5 CLOSER VIEW OF THE COGNITIVE SQUARE

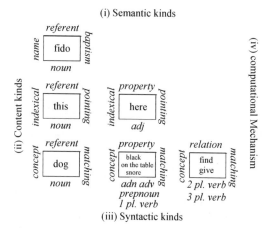

(i) Semantic kinds

(iii) Syntactic kinds

The surfaces inside the rectangles have the following proplet definitions:

4.1.6 PROPLETS INSTANTIATING THE COGNITIVE SQUARE OF DBS

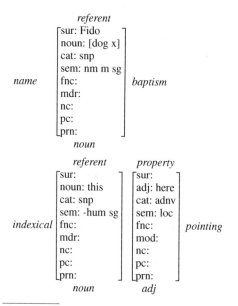

---

[1] Examples in precomputational work are (i) matching for concepts but without the type-token relation and its computational implementation based on content and pattern proplets, (ii) pointing for indexicals but without an on-board orientation system (OBOS), and (iii) baptism but without the named referent as the core value for use in the speak and the hear mode.

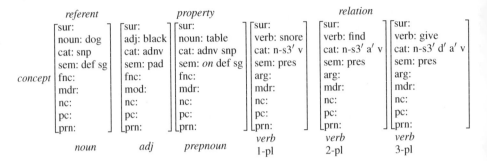

In a proplet, the Semantic kind (a) *referent* is limited to the core attribute noun with the cat values snp and pnp, (b) *property* is limited to the core attribute noun with the cat value adnv, and verb with the cat values adnv and for intransitive, and (c) *relation* is limited to verbs characterized by their cat value as transitive.

The Content kinds *name, indexical,* and *concept* are specified by the core value. In names, the corresponding computational Mechanism of *baptism* is implemented by inserting a 'named referent' as the core value into a lexical proplet, in indexicals by *pointing* at a STAR value of the onboard orientation system, and in concepts by computational type-token *matching.*

The cognitive square[2] of DBS is empirically important because (i) figurative use is restricted to concepts, i.e., the bottom row in 4.1.4–4.1.6, and (ii) reference is restricted to nouns, i.e., the left-most column. Thus only concept nouns may be used both figuratively and as referents, while indexical properties like here and now may not be used as either, and names and indexical nouns like this only as referents.

4.2 Restriction of Figurative Use to Concepts

To show the restriction of figurative use to the Content kind concept let us go systematically through the three Semantic kinds:

4.2.1 THREE CONTENT KINDS FOR THE SEMANTIC KIND *REFERENT*

concept	indexical	name
sur: noun: animal cat: sn ...	sur: noun: pro2 cat: sp2 ...	sur: tom noun: [person x] cat: snp ...

The three Content kinds of the Semantic kind *referent* all have literal use, but only the concepts allow figurative use.

---

[2] 'Triangle' would be appropriate as well, but the term "cognitive triangle" is already used by the cognitive behavioral therapists (CBT). Earlier it was used by Ogden&Richards (1923) for their "Semiotic Triangle." The term 'square' is well suited to express the orthogonal relation between the Syntactic_kinds-Semantic_kinds and the Content_kinds-computational_Mechanisms.

Next consider the Semantic kind *property*, which occurs as the Content kinds (i) *concept* and (ii) *indexical*, but not as *name*. Property proplets of the content kind concept may have the core attributes adj, noun, or verb if it is 1-place (4.1.6). If they have the core value adnv, they may be (a) elementary (fast) or phrasal (in the park), and (b) adnominal (tree in the park) or adverbial (walk in the park).[3]

4.2.2 Two Content kinds for the Semantic kind *PROPERTY*

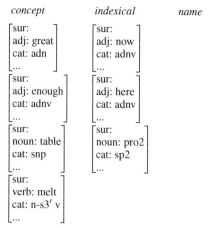

Of the Semantic kind *property*, only the concepts may be used nonliterally.[4]

The Grammatical kind transitive verb with its single Semantic kind *relation* exists as the Content kind *concept*, but not as *indexical* or *name*:

4.2.3 One Content kind for the Semantic kind *RELATION*

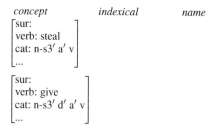

Being concepts, transitive verbs have literal and nonliteral use.

---

[3] Phrasal modifiers, called prepnouns in DBS, are derived from a referent by means of a preposition or an affix, depending on the language. Therefore, a prepnoun like in the park refers by means of park, in contradistinction to the other modifiers, e.g., elementary fast or intransitive snore, which do not refer.

[4] For a nonliteral use of *on the table* see CC 9.2.4 and of *melt* CC 9.5.1.

4.3 Additional Constraint on Figurative se

The restriction of figurative use to concepts is constrained further by the condition that the literal term and its figurative counterpart must be grammatically equivalent:

4.3.1 INVARIANCE CONSTRAINT

A figurative use and its literal counterpart must be of the same Syntactic and Semantic kind.

Thus, one cannot use a 1-place verb like bark to refer figuratively to a dog unless bark is nominalized, as in the little barker (i.e., by turning the property of intransitive bark into the referent barker, sleep into sleeper, stink into stinker, etc.). Similarly for the adj fat, which for figurative use must be nominalized, as in the old fatso. Functionally, the constraint helps the hearer to find the literal counterpart of a figurative use by reducing the search space.

The systematic examples in CC 9 all satisfy the invariance constraint:

4.3.2 SYNTACTIC-SEMANTIC INVARIANCE OF FIGURATIVE USE

Semantic kind	Syntactic kind	nonliteral use	literal counterpart	in CC
referent	noun	animal	dog	9.1.2
property	prepnoun	on the table	on the orange crate	9.2.1
property	adn	great	greater than average	9.6.3
property	adv	enough	more than enough	9.6.6
property	intransitive verb	melt	disappear	9.5.1
relation	transitive verb	steal	take over	9.4.2

The Semantic kind *property* has several Syntactic kinds, while each Syntactic kind, e.g., prepnoun, has only one Semantic kind, i.e., *property*, regardless of whether it is used literally or figuratively. The other two Semantic kinds, i.e., *referent* and *relation*, each have only a single syntactic counterpart.

As an example of using all three Semantic kinds figuratively consider the following description of a dog contorting itself catching a frisbee in mid air:

4.3.3 EXAMPLE USING ALL THREE SEMANTIC KINDS FIGURATIVELY

The animal flew acrobatically towards the disc.

The content obeys the invariance constraint: literal dog and figurative animal are both singular nouns, literal jumped and figurative flew are both finite verbs in the indicative past, literal in a spectacular gymnastic feat and figurative acrobatically are both adverbials (one phrasal, the other elementary), and literal frisbee and figurative disc are both singular nouns. For successful communication, the hearer-reader must relate figurative animal to literal dog and figurative disc to literal frisbee. The relation flew and the property acrobatically, in contrast, do not refer, but establish a relation between referents, or modify a referent, a relation, or a property.

4.4 Declarative Specification of Concepts for Recognition

Concepts are the only Semantic kind which interacts directly with the agent's cognition-external environment. The interaction consists of matching between (i) raw data provided by sensors and activators of the agent's interface component and (ii) concept types provided by the agent's memory.[5]

Consider the rule for the recognition of a color:

4.4.1 DECLARATIVE SPECIFICATION FOR RECOGNITION OF blue

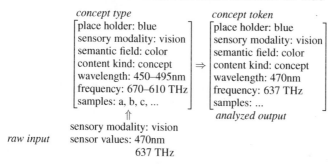

The raw input data 470nm and 637 THz are provided by the agent's interface component and recognized as the color blue because they fall into the type's wavelength interval of 450–495nm and frequency interval of 670–610 THz. The analyzed output token results from replacing the wavelength and frequency intervals of the type with the raw data measurements of the input.

The place holder value of the recognized token, i.e., the letter sequence b l u e, is used for lookup of the lexical proplet which contains the place holder as its core value (CC 1.6.3):

4.4.2 PLACE HOLDER VALUE OF CONCEPT USED FOR LEXICAL LOOKUP

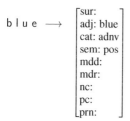

Like the concept type, the proplet is retrieved from the artificial agent's memory (on-board database). Computationally, the lookup is based on string search (Knuth et al. 1977) in combination with a trie structure (Briandais 1959, Fredkin 1961).

The language counterpart to the recognition of nonlanguage concepts is the interpretation of language-dependent surfaces. As an example, consider the DBS robot's

---

[5] From a theory of science point of view, computational pattern matching based on the type-token relation constitutes a fruitful interaction between the humanities and the sciences.

recognition of the letter sequence blue by matching raw visual input data with letter patterns as shape types, resulting in a surface token:

4.4.3 RECOGNITION OF A WRITTEN LANGUAGE SURFACE

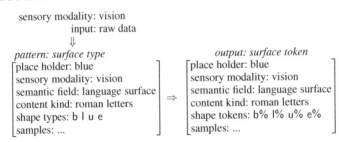

Raw data input is matched by the shape types of the letters b l u e. The output replaces the matching shape types with the shape tokens b% l% u% e%[6] to record such accidental properties as the font, size, color, etc. in the sensory medium of print, and pronunciation, pitch, speed, loudness, etc. in the sensory medium of speech. The shape types are used for matching the raw data and the place holder for look-up of the lexical definition. For developing the linguistic side of automatic word form recognition, the type-token matching of raw data in different media may be cut short temporarily by typing the letters of the place holder directly into a standard computer.

4.5 Declarative Specification of Concepts for Action

The action counterpart to the recognition of nonlanguage concepts is their cognition-external realization as raw data. It consists in adapting a type to the agent's purpose as a token which is passed to the appropriate actuator. As an example, consider a cuttlefish Metasepia Pfefferi turning on the color blue:

4.5.1 RULE FOR PRODUCING THE COLOR blue

concept type
$$\begin{bmatrix} \text{place holder: blue} \\ \text{sensory modality: visual display} \\ \text{semantic field: color} \\ \text{content kind: concept} \\ \text{wavelength: 450–495nm} \\ \text{frequency: 670–610 THz} \\ \text{samples: a, b, c, ...} \end{bmatrix} \Rightarrow$$

concept token
$$\begin{bmatrix} \text{place holder: blue} \\ \text{sensory modality: visual display} \\ \text{semantic field: color} \\ \text{content kind: concept} \\ \text{wavelength: 470nm} \\ \text{frequency: 637 THz} \\ \text{samples: ...} \end{bmatrix}$$
⇓

sensory modality: vision
actuator values: 470nm   *raw output*
637 THz

The type is adapted into a token by replacing the wavelength interval of 450–495nm and frequency interval of 670–610 THz with the agent-selected values of 470nm and

637 THz. In the cuttlefish, these values are realized by natural actuators for color control (chromatophores) as raw data.

The language counterpart to a nonlanguage action is the realization of a language-dependent surface in a medium of choice. As an example, consider the DBS robot's production of the surface blue as raw data in vision:

4.5.2 REALIZING LETTER TOKENS AS RAW DATA IN VISION MEDIUM

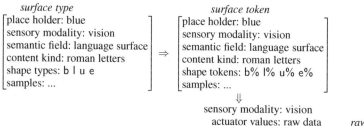

The input to the actuator consists of a sequence of shape tokens representing roman letters. The output replaces the shape tokens, here b% l% u% e%, with matching raw data, for example, pixels on a computer screen.

4.6 Indirect Grounding of Indexicals and Names

In DBS, the second computational Mechanism is *indexicals* pointing at STAR values of the agent's on-board orientation system (OBOS) and is as such cognition-internal. However, because the STAR values originate as concept recognitions, past or present (CC 7, 8), indexicals rely indirectly on the Mechanism of computational pattern matching. More specifically, the indexical pro1 points at the A value of the STAR, pro2 at the R value, pro3 at the 3rd value, here at the S value, and now at the T value (with the S and T values nominalized).

The third computational Mechanism is *names*; it relies on an act of baptism, which inserts the 'named referent' as the core value into a lexical name proplet (CTGR'17). Because the named referent originates as concept cognition, names – like indexicals – rely indirectly on the first computational Mechanism of *concepts*, i.e., computational pattern matching.

After working out the basic functioning of the computational Mechanism for the recognition and action of certain concepts in a codesigned but real environment, more concepts of the same kind (semantic field) may be added routinely, as shown by the following example:

---

[6] The letter shapes are represented by the letters themselves, e.g., e (type) and e% (token).

4.6.1 SIMILARITY AND DIFFERENCE BETWEEN COLOR CONCEPT TYPES

place holder: red sensory modality: vision semantic field: color content kind: concept wavelength: 700-635 nm frequency: 430-480 THz samples: a, b, c, ...	place holder: green sensory modality: vision semantic field: color content kind: concept wavelength:495-570 nm frequency: 526-606 THz samples: a′, b′, c′, ...	place holder: blue sensory modality: vision semantic field: color content kind: concept wavelength: 490-450 nm frequency: 610-670 THz samples: a″, b″, c″, ...

Once the recognition and action side of these concepts is working as intended, more colors may be easily added as an efficient, transparent upscaling.[7]

Similarly for geometric forms: once the concepts of square (CC 1.3.2) and rectangle work as intended, more two-dimensional forms, such as triangle, heptagon, hexagon, and rhombus, may be added routinely. After implementing the concept pick including the associated hand-eye coordination and the semantic relation of object\predicate (CC 2.5.1, 2), the robot should be able to execute language-based requests like Pick the blue square or Pick the green rectangle correctly from a set of items in its task environment.

4.7 Conclusion

In data-driven agent-based DBS, recognition and action must be *grounded* in the form of a computational interaction between raw data and a robot's digital cognition. For practical reasons, grounding may be temporarily suspended by the shortcut of typing place holder values into a standard computer's key board and displaying output on the screen. This allows systematic upscaling of an artificial cognition even today, yet prepares for integrating operational core values for referents, properties, and relations when they become available in robotics.

Computational upscaling has two basic aspects: the *declarative specification* and the *procedural implementation*. A declarative specification must be both, (i) tolerably readable by humans and (ii) easily translatable into a general purpose programming language like Lisp, C, Java, or Perl. From a humanities point of view, a declarative specification must represent the necessary properties of a software solution by omitting the accidental properties which distinguish the individual programming languages and make them difficult to read. Methodologically, a procedural implementation is important because it complements a declarative specification with automatic verification, which supports systematic incremental upscaling.

---

[7] Set-theoretically, the colors red, green, and blue are (i) disjunct and (ii) subsets of color. This structure is inherent in the color concepts, but regardless of being true, it is neither the only nor the predominant aspect of their meaning: knowing that red and green are disjunct, for example, is not sufficient for naming these colors correctly.

5. Comparison of Coordination and Gapping

The most basic distinction in the classical semantic relations of structure is between (i) functor-argument and (ii) coordination. Functor-argument connects different kinds of contents, namely (a) referent/relation (subject/predicate), (b) referent\relation (object\predicate), and (c) property|referent, property|relation, as well as property|property (modifier|modified). Coordination connects[1] the same kinds of content, namely (a) referent−referent, (b) property−property, and (c) relation−relation (conjunct−conjunct), at the elementary, phrasal, and clausal level of grammatical complexity (5.1–5.6). Semantically related but syntactically different are the subject, predicate, and object *gapping* constructions (5.7–5.9).

Examples representing the constructions are systematically analyzed as (i) contents defined as sets of proplets connected by address and as (ii) graphical presentations of the semantic relations of structure. These brief but concise manners of analysis bring out the syntactic-semantic differences between coordination and gapping in general as well as the differences within the coordination constructions and within the gapping constructions in particular.

5.1 Coordination of Elementary Adnominals

The distinction between functor-argument and coordination is established in the data structure of proplets, defined as non-recursive feature structures with ordered attributes. The continuation attributes of functor-argument are fnc, arg, and mdd while those of coordination are nc (next conjunct) and pc (previous conjunct).

An example of a modifier−modifier coordination at the elementary level of grammatical complexity is tall, cool, black, new in the following content:

5.1.1 CONTENT OF The tall, cool, black, new building collapsed.

sur:	sur:	sur:	sur:	sur:	sur:
noun: **building**	adj: **tall**	adj: **cool**	adj: **black**	adj: **new**	verb: **collapse**
cat: snp	cat: adn	cat: adn	cat: adn	cat: adn	cat: #n′ decl
sem: def sg	sem: pad	sem: pad	sem: pad	sem: pad	sem: ind past
fnc: **collapse**	mdd: **building**	mdd:	mdd:	mdd:	arg: **building**
mdr: **tall**	mdr:	mdr:	mdr:	mdr:	mdr:
nc:	nc: **cool**	nc: **black**	nc: **new**	nc:	nc:
pc:	pc:	pc: **tall**	pc: **cool**	pc: **black**	pc:
prn: 23	prn: 23	prn: 23	prn: 23	prn: 23	prn: 23

---

[1] For an overview of exceptions to the grammatical equality of conjuncts and proposals for their resolution see Bruening and Al Khalaf (2020).

© The Author(s), under exclusive license to Springer Nature Switzerland AG 2023
R. Hausser, *Ontology of Communication*, https://doi.org/10.1007/978-3-031-22739-4_5

This content is a set (order-free) of self-contained proplets with (i) the *core values* of the attributes noun, adj, and verb, (ii) the *continuation values* of the attributes fnc, arg, mdr, mdd, nc, and pc, and (iii) the shared prn value, here 23.

The modification relation between the adn coordination *tall cool black new* and the noun *building* is tall|building. It is coded by the features [mdr: tall] of *building* and [mdd: building] of the initial conjunct *tall*. In the noninitial conjuncts, in contrast, the mdd attributes have no value; if needed, it can be retrieved from the initial conjunct via the pc connections (NLC 8).

The semantic relations coded in the content 5.1.1 may be shown as the following graph, whereby the different slashes /, |, and — represent the subject/predicate, modifier|modified and conjunct—conjunct relations.[2]

5.1.2 GRAPHICAL PRESENTATION OF SEMANTIC RELATIONS IN 5.1.1

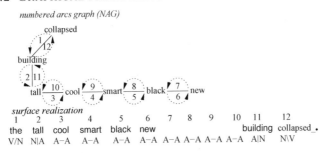

As shown in the *surface realization*, The is realized from the goal proplet in arc 1, tall from the goal proplet in arc 2, cool in arc 3, smart in arc 4, black in arc 5, and new in arc 6. After empty return via the arcs 7-10, building is realized from the goal proplet of arc 11, and collapsed_. of arc 12. The navigation operations are named after the semantic relations they traverse and are shown in the bottom line, beneath the surface. The direction of the traversals is specified by the arrows of the arcs listed by number in the top line of the surface realization.

5.2 Coordination of Phrasal Adnominal Modifiers

In English, phrasal modifiers (prepnouns) consist of a preposition and a noun, e.g., in the water (noun concept), in here (noun indexical), or in Paris (noun name). In contrast to elementary modifiers, which may morphologically distinguish between adnominal and adverbial use, as in beautiful woman vs. sang beautifully, no such distinction exists in phrasal modifiers. Thus, in the water may be used adnominally (5.2.1) and adverbially (5.3.1). Also, while elementary modifiers in adnominal use precede the modified noun, phrasal modifiers follow. Consider the content of The man in the water for days without a lifejacket survived:

5.2.1 SUBJECT MODIFIED BY PHRASAL MODIFIER COORDINATION

sur:	sur:	sur:	sur:	sur:
noun: **man**	noun: **water**	noun: **day**	noun: **life jacket**	verb: **survive**
cat: snp	cat: adnv	cat: adnv	cat: adnv	cat: #n′ decl
sem: def sg	sem: *in* def sg	sem: *for* indef pl	sem: *without* indef sg	sem: ind past
fnc: **survive**	fnc:	fnc:	mdd:	arg: **man**
mdr: **water**	mdd: **man**	mdr:	mdr:	mdr:
nc:	nc: **day**	nc: **life jacket**	nc:	nc:
pc:	pc:	pc: **water**	pc: **day**	pc:
prn: 26	prn: 26	prn: 26	prn: 26	prn: 26

By time-linear absorption of the determiner into the preposition and the noun into the preposition-determiner combination (CLaTR 7.2.5), a prepnoun is represented as a single proplet, like a case-marked locative in classical Latin. The core attribute of phrasal modifiers is noun, but their semantic role as modifiers is specified by the cat value adnv, for adjective with adnominal and adverbial use. In each conjunct, the preposition (in *italics*) is preserved for the speak mode as the first sem value.

Phrasal conjuncts and phrasal modifiers have adnominal as well as adverbial use. The uses are distinguished by word order in conjuncts (5.2.1 vs. 5.3.1), but create an ambiguity between an adnominal (TExer 1.5.3) and an adverbial (TExer 1.5.4) reading in modifiers. The repetition of phrasal modifiers requires the same kind, whereas no such restriction holds for the repetition of phrasal conjuncts. For example, in the modifier repetition on the table (locational) under the tree (locational) in the garden (locational) the modifiers are all of the same modality (TExer 5.1), but in the conjunct repetition in the water (locational) for days (temporal) without a life jacket (manner) the modalities of the conjuncts are all different.

5.2.2 GRAPHICAL PRESENTATION OF SEMANTIC RELATIONS IN 5.2.1

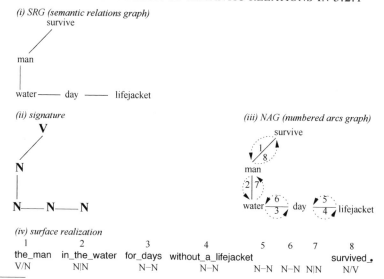

(i) SRG (semantic relations graph)

survive

man

water —— day —— lifejacket

(ii) signature

V

N

N —— N —— N

(iii) NAG (numbered arcs graph)

survive

man

water day lifejacket

(iv) surface realization

1	2	3	4	5	6	7	8
the_man	in_the_water	for_days	without_a_lifejacket				survived .
V/N	N|N	N–N	N–N	N–N	N–N	N|N	N/V

---

[2] In substitution-based linguistics (PSG), there is some agreement that the flat concatenation of coordination is a difficulty for constituent structure: Ross (1967), Dik (1968), Goldsmith (1985), Sag,

As in 5.1.2, the modifier|modified relation between the phrasal modifier coordination and the modified noun is traversed in arcs 2 (downward) and 7 (upward). Fulfillment of the continuity condition (NLC 3.6.5) as the think-speak mode counterpart to (and the source of) the time-linear derivation order in the hear mode is clearly shown in the bottom line of the *(iv) surface realization*, i.e., the goal proplet of operation n equals the start proplet of operation n+1.

5.3 Coordination of Phrasal Adverbial Modifiers

The distinction between the adnominal and the adverbial use of one and the same phrasal modifier coordination is located in the connection between the modified and the initial conjunct, e.g., between man and in the water in 5.2.1 (adnominal), and between survived and in the water in 5.3.2 (adverbial).

5.3.1 PREDICATE MODIFIED BY PHRASAL MODIFIER CONJUNCTION

⎡sur: ⎤	⎡sur: ⎤	⎡sur: ⎤	⎡sur: ⎤	⎡sur: ⎤
noun: **man**	verb: **survive**	noun: **water**	noun: **days**	noun: **lifejacket**
cat: snp	cat: #n' v	cat: snp	cat: snp	cat: snp
sem: def sg	sem: ind past	sem: *in* def sg	sem: *for* def sg	sem: *without* indef sg
fnc: **survive**	arg: **man**	fnc: **survive**	mdd:	mdd:
mdr:	mdr: **water**	mdr:	mdr:	mdr:
nc:	nc:	nc: **days**	nc: **lifejacket**	nc:
pc:	pc:	pc:	pc: **water**	pc: **day**
⎣prn: 25 ⎦	⎣prn: 25 ⎦	⎣prn: 25 ⎦	⎣prn: 25 ⎦	⎣prn: 25 ⎦

This presentation of the content as a set of proplets is complemented by the standard representation as a semantic relations graph:

5.3.2 GRAPHICAL PRESENTATION OF SEMANTIC RELATIONS IN 5.3.1

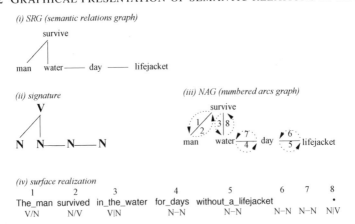

(i) SRG (semantic relations graph)

survive

man   water——— day ——— lifejacket

(ii) signature

V

N N——N——N

(iii) NAG (numbered arcs graph)

survive

man   water 4 day 5 lifejacket

(iv) surface realization

1	2	3	4	5	6	7	8
The_man	survived	in_the_water	for_days	without_a_lifejacket			•
V/N	N/V	V|N	N–N	N–N	N–N	N–N	N|V

The modifier|modified relation between the phrasal modifier coordination and the modified verb is traversed in arcs 3 (downward) and 8 (upward).[3]

Gazdar, Wasow, and Weisler (1985), Lakoff (1986), Bayer (1996), Osborne (2006), and others.

5.4 Coordination of Elementary Nouns as Subject

From the coordination of modifiers in 5.1–5.3, we turn to the coordination of arguments, i.e., subject or object. The following coordination of names is coded via the nc and pc values and used as the grammatical subject:

5.4.1 NOUN COORDINATION SERVING AS SUBJECT

sur: fido	sur: tucker	sur: buster	sur:	sur:
noun: [dog x]	noun: [dog y]	noun: [dog z]	verb: **snore**	adj: **loud**
cat: snp	cat: snp	cat: snp	cat: #n' decl	cat: adv
sem: nm m	sem: nm m	sem: *and* nm m	sem: ind past	sem: pad
fnc: **snore**	fnc:	fnc:	arg: [dog x]	mdd: **snore**
mdr:	mdr:	mdr:	mdr: **loud**	mdr:
nc: [dog y]	nc: [dog z]	nc:	nc:	nc:
pc:	pc: [dog x]	pc: [dog y]	pc:	pc:
prn: 18	prn: 18	prn: 18	prn: 18	prn: 18

In contrast to elementary (5.1.1) and phrasal (5.2.1) coordinations serving as modifiers, coordinations of nouns serving as argument require the prefinal conjunction *and*,[4] coded as the initial sem value of the final conjunct.

5.4.2 GRAPHICAL PRESENTATION OF SEMANTIC RELATIONS IN 5.4.1

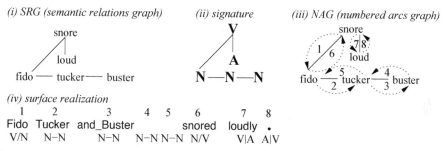

(i) SRG (semantic relations graph)   *(ii) signature*   *(iii) NAG (numbered arcs graph)*

(iv) surface realization

1	2	3	4	5	6	7	8
Fido	Tucker	and_Buster			snored	loudly	.
V/N	N–N	N–N	N–N	N–N	N/V	V/A	A/V

The semantic relation between a noun coordination and the predicate is based on the standard subject/predicate or object\predicate relation, using the initial conjunct (here in arcs 1 and 6). In the content 5.4.1, this relation is coded by the [fnc: snore] feature of the initial conjunct *fido* and the [arg: [dog x]] feature of the verb *snore*. In the noninitial conjuncts, the fnc attributes have no value; if needed, it can be retrieved from the initial conjunct via the pc connections (NLC 8.3.3 ff).

5.5 Intra- and Extrapropositional Verb Coordination

While adn and noun coordinations are intrapropositional, verb coordination may also be extrapropositional. This is because DBS represents a proposition by its top verb,

---

[3] Comparison with TExer 5.1.12 shows the semantic difference between the intrapropositional repetition of modification vs. coordination.

[4] For the graph analysis and for the complete sequence of hear mode operations see TExer 3.6.

whereby the complete content may be reconstructed by navigating along the continuation values. In a text or dialogue, the traversal of the first proposition begins with the top verb, continues along the continuation values, returns to the current top verb, and continues to the top verb of the next proposition by extrapropositional coordination (5.6.2). A top verb with an empty nc slot concludes an extrapropositional traversal.

Intra- and extrapropositional verb coordinations may combine as follows:

5.5.1 INTRA- IN EXTRAPROPOSITIONAL VERB COORDINATION

Julia slept. Bob bought, peeled, and ate an apple. Fido snored.

[prn: n]   [prn: n+1]   [prn: n+2]

The critical transition is from the intrapropositional coordination of [prn: n+1] to the next proposition [prn: n+2] by means of an extrapropositional verb−verb coordination. The following solutions have been proposed:

5.5.2 ALTERNATIVE NAGs FOR EXTRAPROPOSITIONAL VERB COORDINATION

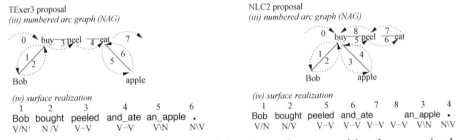

TExer3 proposal
(iii) numbered arc graph (NAG)

(iv) surface realization

1	2	3	4	5	6
Bob	bought	peeled	and_ate	an_apple	.
V/N'	N /V	V–V	V–V	V\N	N\V

NLC2 proposal
(iii) numbered arc graph (NAG)

(iv) surface realization

1	2	5	6	7	8	3	4
Bob	bought	peeled	and_ate			an_apple	.
V/N	N/V	V–V	V–V	V–V	V–V	V\N	N\V

The (obsolete) NLC2 analysis on the right takes an intrapropositional perspective by treating subject (5.4), predicate (NLC2 8.3.4), object (NLC2 8.2.7), and modifier (5.1–5.2) coordinations alike. The initial conjunct *buy* is (a) the representative of the proposition as the carrier of the syntactic mood value, (b) the point of extrapropositional entrance, and (c) the point of extrapropositional exit.

If there is only a single top verb, which is usually the case, this is easily fulfilled. However, if there are several verbs of equal rank, e.g., the intrapropositional verb coordination in n+1 of 5.5.1, the NLC2 proposal would have to allow two values in the nc slot of the initial conjunct *buy*, one for the intrapropositional conjunct peel, the other for the extrasentential conjunct snore.

The TExer proposal avoids this complication by implementing verb conjunctions in the forward direction only, leaving a possible backward navigation to the following inference:

5.5.3 BACKWARD NAVIGATION INFERENCE FOR VERBAL CONJUNCTION

$$\begin{bmatrix} \text{verb: } \beta \\ \text{pc: } \alpha \\ \text{prn: } n+1 \end{bmatrix} \Rightarrow \begin{bmatrix} \text{verb: } \alpha \\ \text{nc: } \beta \\ \text{prn: } n \end{bmatrix}$$

In summary, while almost all functor-argument and coordination relations are implemented bidirectionally, the backward traversal of verbal conjunctions in the speak mode is treated by inference (5.5.3) instead of a routinely provided V←V operation. This is because a return traversal in verbal coordination (i) is not necessary, as demonstrated by the TExer solution shown in 5.5.2, (ii) may therefore only be used when rhetorically desired, as when telling a story starting from the end, and (iii) requires specific operators like before that, appropriately specified by the inference.

5.6 Extrasentential Coordination

The most common extrasentential connection between sentences in a text is extrasentential coordination (parataxis):

5.6.1 CONTENT OF Mary slept. Fido snored.

$$
\begin{bmatrix}
\text{sur: mary} \\
\text{noun: [person x]} \\
\text{cat: snp} \\
\text{sem: nm f} \\
\text{fnc: sleep} \\
\text{mdr:} \\
\text{nc:} \\
\text{pc:} \\
\text{prn: 17}
\end{bmatrix}
\begin{bmatrix}
\text{sur:} \\
\text{verb: sleep} \\
\text{cat: #ns3 decl} \\
\text{sem: ind past} \\
\text{arg: [person x]} \\
\text{mdr:} \\
\text{nc: (snore 18)} \\
\text{pc:} \\
\text{prn: 17}
\end{bmatrix}
\begin{bmatrix}
\text{sur:} \\
\text{verb: snore} \\
\text{cat: #ns3 decl} \\
\text{sem: ind past} \\
\text{arg: [dog y]} \\
\text{mdr:} \\
\text{nc:} \\
\text{pc: (sleep 17)} \\
\text{prn: 18}
\end{bmatrix}
\begin{bmatrix}
\text{sur: fido} \\
\text{noun: [dog y]} \\
\text{cat: snp} \\
\text{sem: nm m} \\
\text{fnc: snore} \\
\text{mdr:} \\
\text{nc:} \\
\text{pc:} \\
\text{prn: 18}
\end{bmatrix}
$$

The values in the nc and pc slots are the extrapropositional addresses (snore 18) and (sleep 17). The multiple operation applications for simultaneously establishing functor-argument and coordination relations within the proplet set are data-driven, i.e., there is no need for additional software.

The pivot of an extrasentential coordination in the hear mode derivation is the interpunctuation between sentences (Ballmer 1978). The interpunctuation proplet (i) supplies the syntactic mood value to the top verb of the present sentence, (ii) cross-copies with the intervening subject of the next sentence, and (iii) absorbs the next verb, thus becoming the predicate of the next sentence. These steps leave no trace in the content 5.6.1 and in the semantic relations graph:

5.6.2 GRAPHICAL PRESENTATION OF SEMANTIC RELATIONS IN 5.6.1

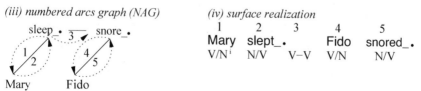

(iii) numbered arcs graph (NAG)

(iv) surface realization

1	2	3	4	5
Mary	slept_.		Fido	snored_.
V/N'	N/V	V–V	V/N	N/V

For the complete declarative specification of an extrasentential coordination see TExer 2.1.

5.7 Quasi Coordination in Subject Gapping

In linguistics, a grammatical construction in which a single *shared* item is in a semantic relation with a sequence of n (n ≥ 1) 'gapped' items is called *gapping*. Basic examples are (i) subject gapping, (ii) predicate gapping, and (iii) object gapping,[5] which have the following pretheoretical structure:

5.7.1 PRETHEORETICAL COMPARISON OF THREE GAPPING KINDS

subject gapping	*predicate gapping*	*object gapping*
Bob buy apple	Bob **buy** apple	Bob buy ∅
∅ peel pear	Jim ∅ pear	Jim peel ∅
and ∅ eat peach	*and* Bill ∅ peach	*and* Bill eat **peach**

The shared item is shown in bold face, while the gapped items are indicated by the gap marker ∅.[6]

The following example shows the content of a subject gapping:

5.7.2 CONTENT OF A SUBJECT GAPPING

Bob bought an apple, ∅ peeled a pear, and ∅ ate a peach.

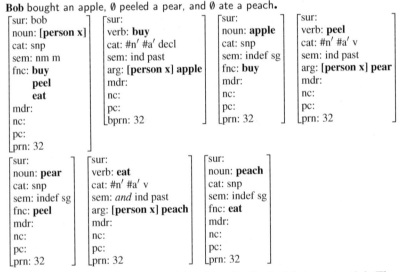

Gapping constructions are intrapropositional, and a fortiori intrasentential. They are only quasi-coordinations because the nc and pc slots are not involved, i.e., they have

---

[5] There seems to be no "modifier gapping" in natural language.

[6] Even in GG, there is some agreement that gapping constructions do not conform to constituent structure: Ross (1967, 1970), Jackendoff (1971), Kuno (1976), Sag (1976), Hankamer (1979), McCawley (1988), Hartmann (2000), Osborne (2006), Johnson (2009), and others.

no intrapropositional values. They resemble nominal and intrapropositional verb coordinations, however, in that they use prefinal *and* and consist of unbounded repetitions of grammatically similar items.

The semantic relations between the shared item *bob* and the gapped items Ø *peel pear* and Ø *eat peach* are run via the gap list in the shared item and the repetition of the shared item's address, here [person x], in the verbs of the gapped items (arg slot, initial position). In this way, the semantic relations of structure are complete in a gapping construction without using the nc and pc slots.

5.7.3 GRAPHICAL PRESENTATION OF SEMANTIC RELATIONS IN 5.7.2

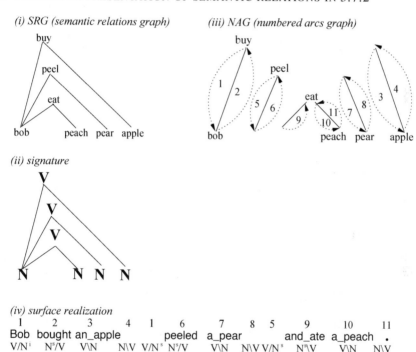

(i) SRG (semantic relations graph)

(iii) NAG (numbered arcs graph)

(ii) signature

(iv) surface realization

1	2	3	4	1	6		7	8	5	9	10	11
Bob	bought	an_apple			peeled	a_pear				and_ate	a_peach	.
V/N^i	N^s/V	V\N		N\V	V/N^s	N^s/V	V\N	N\V	V/N^s	$N^a\!V$	V\N	N\V

The different tilts of the three N/V and N\V relations are solely for visual separation in the graph. The gaps appear as empty traversals. The navigation ends with arc 11. The upward arc 9 does not have a downward counterpart. The arc numbering is breadth-first. The number of operations is even. As a multiple verb construction (5.5), the last verb, here *eat*, is used for the extrapropositional exit (TExer 1.4.8).

The think-speak navigation along the semantic relations between proplets is continuous (Continuity Condition, NLC 3.6.5), as shown by the bottom line of the *surface realization*. This is possible by leaving the control of the gaps in the surface to the lexicalization rules, here arcs 4, 1 and 8, 5. For example, lexnoun realizes the surface of the shared noun proplet *bob* (goal proplet of the V/N^s operations in arcs 1, 1, and 5) if, and only if, its initial fnc value is not yet #-marked.

5.8 Quasi Coordination in Predicate Gapping

The pretheoretical characterization of predicate gapping in 5.7.1 is formally instantiated as the following content:

5.8.1 CONTENT OF A PREDICATE GAPPING

Bob **bought** an apple, Jim ∅ a pear, and Bill ∅ a peach.

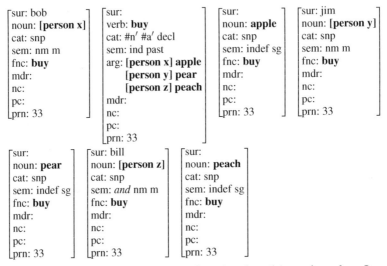

Predicate gapping requires a transitive verb as its shared item, here *buy*. Its arg slot
contains the gap list, here the subject-object pairs *bob apple, jim pear,* and *bill peach.*
The subject and object proplets of the gapped items take buy as their shared fnc value.
The *and* is coded into the initial sem slot of *bill.*

The semantic relations may be shown as the following standard graph:

5.8.2 GRAPHICAL PRESENTATION OF SEMANTIC RELATIONS IN 5.8.1

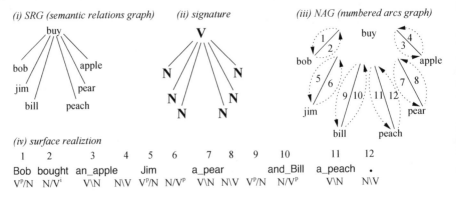

The shared predicate relates to the subject and object of its initial sentence (arcs 1–4) and of its two gapped items (arcs 5–8 and 9–12).

5.9 Quasi Coordination in Object Gapping

Compared to subject and predicate gapping, in which the gaps precede the shared item (filler), object gapping is special in that the filler follows the gaps. Therefore the gap list must be accumulated in an external cache until the filler arrives (strictly time-linear derivation order).

5.9.1 CONTENT OF AN OBJECT GAPPING

Bob bought Ø, Jim peeled Ø, and Bill ate **a peach .**

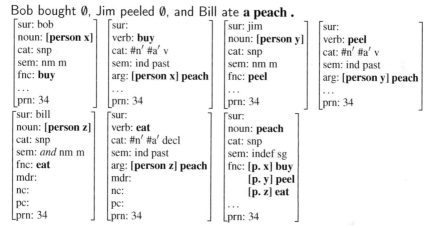

sur: bob		sur:		sur: jim		sur:	
noun: [**person x**]		verb: **buy**		noun: [**person y**]		verb: **peel**	
cat: snp		cat: #n′ #a′ v		cat: snp		cat: #n′ #a′ v	
sem: nm m		sem: ind past		sem: nm m		sem: ind past	
fnc: **buy**		arg: [**person x**] **peach**		fnc: **peel**		arg: [**person y**] **peach**	
. . .		. . .		. . .		. . .	
prn: 34		prn: 34		prn: 34		prn: 34	

sur: bill		sur:		sur:	
noun: [**person z**]		verb: **eat**		noun: **peach**	
cat: snp		cat: #n′ #a′ decl		cat: snp	
sem: *and* nm m		sem: ind past		sem: indef sg	
fnc: **eat**		arg: [**person z**] **peach**		fnc: [**p. x**] **buy**	
mdr:		mdr:		[**p. y**] **peel**	
nc:		nc:		[**p. z**] **eat**	
pc:		pc:		. . .	
prn: 34		prn: 34		prn: 34	

The three verb proplets all take the core value peach as their shared object. The semantic relations of structure may be shown as the following standard graph:

5.9.2 GRAPHICAL PRESENTATION OF SEMANTIC RELATIONS IN 5.9.1

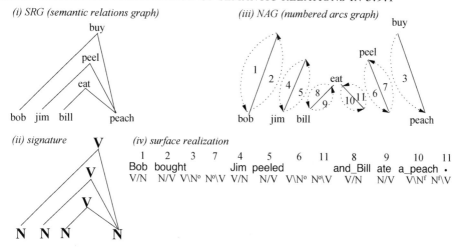

(i) SRG (semantic relations graph)

(ii) signature

(iii) NAG (numbered arcs graph)

(iv) surface realization

The shared object is clearly shown. Just as the graph 5.7.3 for subject gapping is missing a downward arc opposite arc 9, the current graph for object gapping is missing an upward arc opposite arc 3. As a multiverb construction, the last verb, here *eat*, is used for the extrapropositional exit.

5.10 Conclusion

Coordination and gapping have in common that they repeat an unlimited number of similar items. They differ in that the connection between the conjuncts of a coordination is coded by the values of their nc (next conjunct) and pc (previous conjunct) attributes, while no such nc−pc relations exist in gapping constructions.

Instead subject, predicate, and object gapping establish the relation between a single *shared item* and a sequence of repeating *gapped items* by means of (i) a gap list in the shared item and (ii) copies of the shared item's core value in the grammatically appropriate slots of the gapped items. The nc and pc attributes are not used, i.e., they have no intrapropositional values in gapping constructions.

For a complete declarative specification of subject gapping in the hear and speak mode see TExer 5.2, of predicate gapping 5.3, and object gapping 5.4. Related constructions are the unbounded repetition of prepositional phrases and of adnominal clauses, which are analyzed as complete declarative specifications in TExer 5.1 and 5.6, respectively.

6. Are Iterating Slot-Filler Structures Universal?

An example of an iterating slot-filler[1] structure is repeating infinitives, as in John decided to try to persuade Bob to run (6.3). Each infinitive serves as a phrasal object with a slot, e.g., to try *slot*. The slot is filled by another infinitive with a slot, i.e., to persuade Bob *slot*, and so on, except for the last one, which terminates the iteration with an intransitive verb. Similar slot-filler repetitions are iterating object clauses (6.4) and iterating adnominal clauses (6.5).

Another kind of iterating slot-filler or filler-slot structure is known as gapping. For example, Bill bought an apple, peeled a pear, ..., and ate a peach. is called subject gapping. Here the open-ended sequence *slot* peeled a pear, *slot* ate a peach, an so on, takes Bill as the shared subject. The end of the iteration is announced by the function word and in penultimate position. Systematic variants are predicate gapping and object gapping (6.6).

Iterating slot-filler structures may interact with other grammatical constructions. A prime example is a *long distance dependency*, such as Whom did John say that Bill claims that Suzy believes that ... Mary loves?, i.e., the single filler-slot pair Whom and Mary loves? is separated by a slot-filler iteration of object clauses (6.7).

Remarkably, all of these highly conspicuous constructions occur in natural languages which are completely unrelated to English and other European languages, namely in Korean, Tagalog, and Georgian. They are therefore candidates for being universal[2] thought structures.

6.1 Language and Thought

The agent-based data-driven ontology of DBS treats cognitive contents as sets (order-free) of proplets, defined as nonrecursive feature structures with ordered attributes. Proplets are connected into content by the classical semantic relations of structure, coded by address. Language content and thought content are treated alike except that the proplets of a language content have language-dependent sur(face) values which are absent in the proplets of a thought content.

---

[1] McCord (1980) used the slot-filler priniple for *Slot Grammar*.
[2] What exactly constitutes a universal is a difficult question (Harbsmeier 2001).

© The Author(s), under exclusive license to Springer Nature Switzerland AG 2023   57
R. Hausser, *Ontology of Communication*, https://doi.org/10.1007/978-3-031-22739-4_6

A content may use different surfaces from typologically similar languages. For example, The dog found a bone, Der Hund fand einen Knochen, and Le chien a trouvé un os are surfaces in different languages for the same content:

6.1.1 FROM CONTENT TO SPEAK MODE TO HEAR MODE TO CONTENT

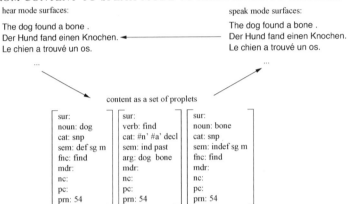

hear mode surfaces:

The dog found a bone .
Der Hund fand einen Knochen. ◄───────────────
Le chien a trouvé un os.

...

speak mode surfaces:

The dog found a bone .
Der Hund fand einen Knochen.
Le chien a trouvé un os.

...

content as a set of proplets

```
⎡ sur:            ⎤  ⎡ sur:             ⎤  ⎡ sur:              ⎤
⎢ noun: dog       ⎥  ⎢ verb: find       ⎥  ⎢ noun: bone        ⎥
⎢ cat: snp        ⎥  ⎢ cat: #n' #a' decl⎥  ⎢ cat: snp          ⎥
⎢ sem: def sg m   ⎥  ⎢ sem: ind past    ⎥  ⎢ sem: indef sg m   ⎥
⎢ fnc: find       ⎥  ⎢ arg: dog bone    ⎥  ⎢ fnc: find         ⎥
⎢ mdr:            ⎥  ⎢ mdr:             ⎥  ⎢ mdr:              ⎥
⎢ nc:             ⎥  ⎢ nc:              ⎥  ⎢ nc:               ⎥
⎢ pc:             ⎥  ⎢ pc:              ⎥  ⎢ pc:               ⎥
⎣ prn: 54         ⎦  ⎣ prn: 54          ⎦  ⎣ prn: 54           ⎦
```

In language communication, the hear mode is provided with a sequence of lexical proplets by automatic word form recognition. Connected into content, they may be used as input to the speak mode for surface production as output (repeated hearsay).

As the computational data structure of DBS, a proplet encodes all lexical and compositional properties of a word as proplet-internal attribute-value pairs called features. For example, the noun proplets in 6.1.1 have the lexical features [sem: def sg] and [sem: indef sg] for definiteness and number, and the verb proplet has the lexical features [cat: #n' #a' decl] for valency and syntactic mood, and [sem: past ind] for tense and verbal mood. The syntactic-semantic distinction between subject and object is coded inside the verb proplet by the value order in the feature [arg: dog bone]. The subject proplet *dog* and the object proplet *bone* are connected to the *find* proplet by their respective [fnc: find] features.

Consider the explicit hear mode derivation of the German example:

6.1.2 SURFACE-COMPOSITIONAL TIME-LINEAR DERIVATION

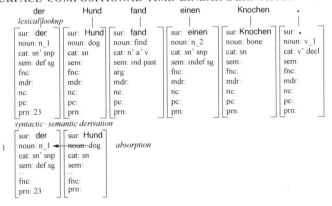

```
            der          Hund         fand          einen         Knochen         .
lexical lookup    |            |             |             |              |
⎡ sur: der    ⎤⎡ sur: Hund ⎤⎡ sur: fand    ⎤⎡ sur: einen  ⎤⎡ sur: Knochen⎤⎡ sur: .      ⎤
⎢ noun: n_1   ⎥⎢ noun: dog ⎥⎢ noun: find   ⎥⎢ noun: n_2   ⎥⎢ noun: bone  ⎥⎢ noun: v_1   ⎥
⎢ cat: sn' snp⎥⎢ cat: sn   ⎥⎢ cat: n' a' v ⎥⎢ cat: sn' snp⎥⎢ cat: sn     ⎥⎢ cat: v' decl⎥
⎢ sem: def sg ⎥⎢ sem:      ⎥⎢ sem: ind past⎥⎢ sem: indef sg⎥⎢ sem:       ⎥⎢ sem:        ⎥
⎢ fnc:        ⎥⎢ fnc:      ⎥⎢ arg:         ⎥⎢ fnc:        ⎥⎢ fnc:        ⎥⎢ fnc:        ⎥
⎢ mdr:        ⎥⎢ mdr:      ⎥⎢ mdr:         ⎥⎢ mdr:        ⎥⎢ mdr:        ⎥⎢ mdr:        ⎥
⎢ nc:         ⎥⎢ nc:       ⎥⎢ nc:          ⎥⎢ nc:         ⎥⎢ nc:         ⎥⎢ nc:         ⎥
⎢ pc:         ⎥⎢ pc:       ⎥⎢ pc:          ⎥⎢ pc:         ⎥⎢ pc:         ⎥⎢ pc:         ⎥
⎣ prn: 23     ⎦⎣ prn:      ⎦⎣ prn:         ⎦⎣ prn:        ⎦⎣ prn:        ⎦⎣ prn:        ⎦
syntactic–semantic derivation
⎡ sur: der    ⎤⎡ sur: Hund ⎤
⎢ noun: n_1 ◄─┤─noun: dog  ⎥  absorption
1 ⎢ cat: sn' snp⎥⎢ cat: sn  ⎥
⎢ sem: def sg ⎥⎢ sem:      ⎥
⎢ ...         ⎥⎢ ...       ⎥
⎢ fnc:        ⎥⎢ fnc:      ⎥
⎣ prn: 23     ⎦⎣ prn:      ⎦
```

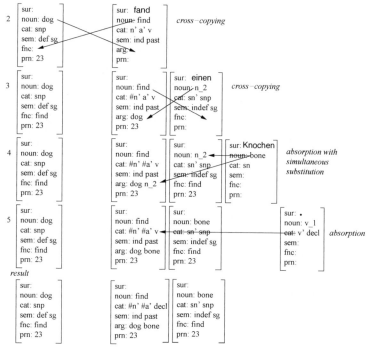

The derivation is *surface-compositional* because each input surface has exactly one lexical proplet representation and there are no lexical proplet representations without a concrete surface. The derivation is *time-linear* as shown by the stair-like addition of one new next word form in each line. The proplets of the function words the and a absorb their respective content words, as shown in lines (1,2) and (4,5), while the . proplet is absorbed into the top verb (5, result).

Contents resulting from hear mode derivations are possible inputs to the think mode operations of (a) selective activation by navigation and (b) inferencing. Either may be mirrored by language-dependent surfaces in the speak mode riding piggyback on sequences of think mode operations:

6.1.3 GRAPHICAL PRESENTATION OF THE SEMANTIC RELATIONS IN 6.1.1

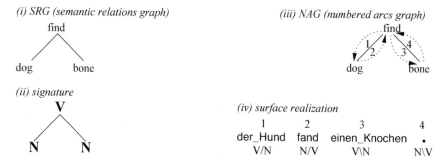

The *(i) SRG* and the *(ii) signature* show the static semantic structure, here subject/

predicate and object\predicate, whereby the nodes in the *SRG* represent the core values, and in the *signature* the core attributes, of the proplets in the content. The *(iii) NAG* and the *(iv) surface realization*, in contrast, show the dynamic aspect of the think mode which activates content by a navigation for inferencing and for the speak mode realization of language-dependent surfaces.

The content 6.1.1, the hear mode derivation 6.1.2, and the speak mode derivation 6.1.3 combine into the following cycle of natural language communication in DBS (shown for English surfaces):

6.1.4 CYCLE OF NATURAL LANGUAGE COMMUNICATION

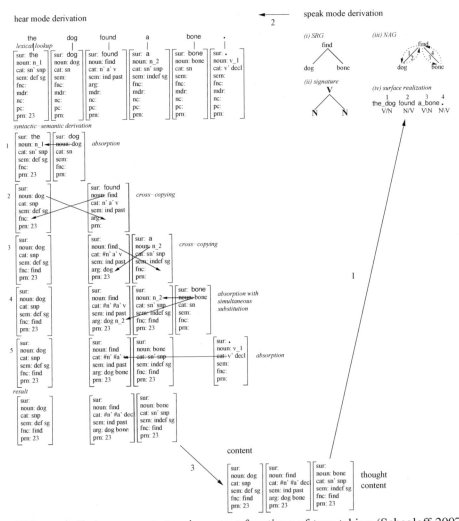

This constellation supports two important functions of turn taking (Schegloff 2007): (i) transfer of a content from the speaker to the hearer (two agents, arrow sequence

1, 3) and (ii) repeated hearsay, i.e., interpreting content in the hear mode and reproducing it in the speak mode (one agent, arrow sequence 3, 1).

6.2 Slot-Filler Iteration

There is comparatively little argument that, roughly speaking, all natural languages distinguish between the word kinds noun, verb, and adj, the syntactic moods declarative, interrogative, and imperative, the verbal moods indicative and subjunctive, the tenses present, past, and future, the semantic relations of functor-argument and coordination, the degrees of elementary, phrasal, and clausal grammatical complexity, and the intra- vs. extrapropositional semantic relations of structure, whereby the latter may be intra- or extrasentential. The basic grammatical structures built on these notions may be considered likely to be universal contents.

In contrast, a structure perhaps more promising for being non-universal is the conspicuous iteration of (i) slot-fillers or (ii) filler-slots such as the following:

6.2.1 EXAMPLES OF SLOT-FILLER AND FILLER-SLOT ITERATIONS

1. *Iterating infinitives* (phrasal)
   John decided to try to persuade Bob to run.
2. *Iterating object clauses* (clausal)
   Mary saw that Peter saw that Suzy saw Fido.
3. *Iterating adnominal clauses* (clausal)
   Mary saw the man who loves the woman who feeds the child.
4. *Subject gapping* (phrasal)
   Bob bought an apple, peeled a pear, and ate a peach.
5. *Predicate gapping* (phrasal)
   Bob bought an apple, Jim a pear, and Bill a peach.
6. *Object gapping* (phrasal)
   Bob bought, Jim peeled, and Bill ate the peach.
7. *Object clause iteration with long distance dependency* (clausal)
   Whom did John say that Bill believes that Mary claims that Suzy loves?

In this list, three construction kinds may be distinguished: (i) marked slot-filler repetition using different fillers (1-3), (ii) unmarked filler-slot or slot-filler repetition using the same filler (4-6), and (iii) a single filler-slot relation with an intervening object clause iteration, resulting in a long-distance dependency (7).

6.3 Marked Slot-Filler Repetition in Infinitives

In English, marked slot-filler repetition occurs at the phrasal level as (a) repeating infinitives,[3] and at the clausal level as (b) repeating object clauses and (c) repeating adnominal clauses. The cognitive structure common to all three may be shown abstractly as follows:

---

[3] In HPSG, infinitives are treated as a kind of clause (Sag 1997). It seems, however, that the adnominal use of infinitives, as in The decision to try..., is restricted to nominalized transitive verbs.

6.3.1 SLOT-FILLER REPETITION AS ABSTRACT COGNITIVE STRUCTURE

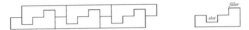

For example, the slot-fillers in an infinitive iteration consist of a sequence of transitive predicates taking another transitive predicate as their second argument, except for the last one. The lower predicate's subject is absent in the surface (slot) but equals implicitly the next higher subject (subject control, e.g., try, promise) or the next higher object (object control, e.g., ask, persuade).

The function word marking the slots in English infinitives is to. The characteristic semantic connections may be illustrated as follows:

6.3.2 MARKED SLOTS IN ITERATING INFINITIVES

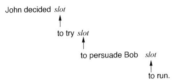

The initial predicate decide_*slot* is filled by to try_*slot* which is filled by to persuade Bob_*slot* which is filled by to run. The last filler terminates the iteration because it does not introduce another slot. Whether an infinitive has subject or object control CLaTR 15.4–15.6) depends on the verb.

Consider the surface-compositional time-linear derivation of John decided to try to persuade Bob to run:

6.3.3 HEAR MODE DERIVATION OF REPEATING INFINITIVES

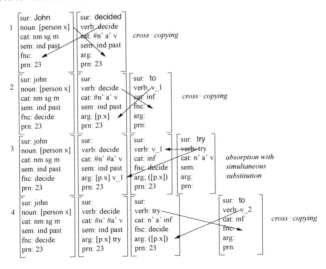

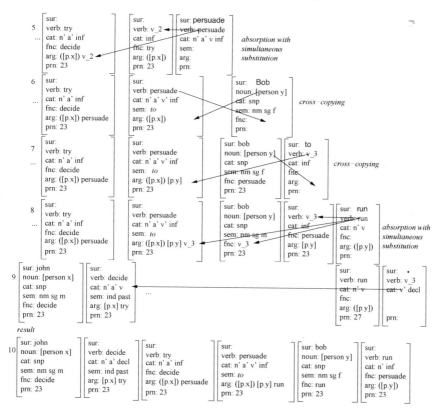

As a surface-compositional approach, DBS connects decide directly with to_try, try directly with to_persuade, and persuade directly with to_run, using the V\V relation intrasententially. The implicit subject and object control of the infinitives is shown explicitly as the values [person x] and [person y], i.e., the named referents of John and Bob (in square brackets).

For the corresponding speak mode, consider the following graph analysis:

6.3.4 GRAPHICAL PRESENTATION OF THE SEMANTIC RELATIONS IN 6.3.3

John decided to try to persuade Bob to run.

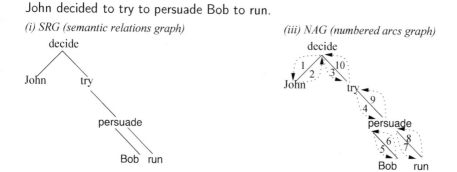

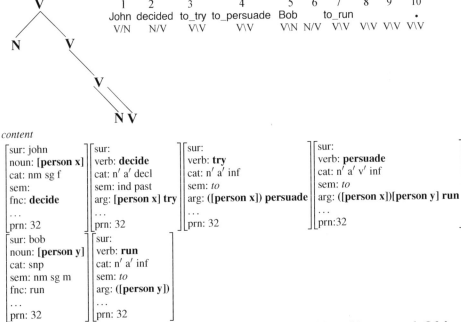

(ii) signature   *(iv) surface realization*

The first two infinitives have subject control, while the third has object control. Of the verbs, run is 1-place, decide and try are 2-place, and persuade is 3-place. The valency positions are marked with ′.

6.4 Marked Slot-Filler Repetition in Object Clauses

The clausal counterpart to phrasal infinitive iteration is the extrapropositional repetition of object sentences, as in the following example:

6.4.1 OBJECT CLAUSE REPETITION

Mary saw that Bill saw that Suzy saw Fido.

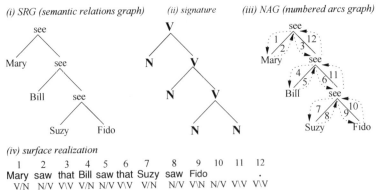

(i) SRG (semantic relations graph)   *(ii) signature*   *(iii) NAG (numbered arcs graph)*

(iv) surface realization

content

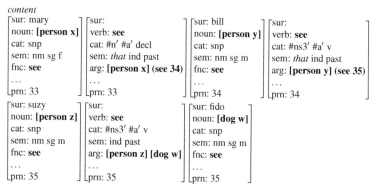

$$\begin{bmatrix} \text{sur: mary} \\ \text{noun: } \textbf{[person x]} \\ \text{cat: snp} \\ \text{sem: nm sg f} \\ \text{fnc: see} \\ \ldots \\ \text{prn: 33} \end{bmatrix} \begin{bmatrix} \text{sur:} \\ \text{verb: see} \\ \text{cat: } \#n' \#a' \text{ decl} \\ \text{sem: } \textit{that} \text{ ind past} \\ \text{arg: } \textbf{[person x]} \textbf{ (see 34)} \\ \ldots \\ \text{prn: 33} \end{bmatrix} \begin{bmatrix} \text{sur: bill} \\ \text{noun: } \textbf{[person y]} \\ \text{cat: snp} \\ \text{sem: nm sg m} \\ \text{fnc: see} \\ \ldots \\ \text{prn: 34} \end{bmatrix} \begin{bmatrix} \text{sur:} \\ \text{verb: see} \\ \text{cat: } \#ns3' \#a' \text{ v} \\ \text{sem: } \textit{that} \text{ ind past} \\ \text{arg: } \textbf{[person y]} \textbf{ (see 35)} \\ \ldots \\ \text{prn: 34} \end{bmatrix}$$

$$\begin{bmatrix} \text{sur: suzy} \\ \text{noun: } \textbf{[person z]} \\ \text{cat: snp} \\ \text{sem: nm sg m} \\ \text{fnc: see} \\ \ldots \\ \text{prn: 35} \end{bmatrix} \begin{bmatrix} \text{sur:} \\ \text{verb: see} \\ \text{cat: } \#ns3' \#a' \text{ v} \\ \text{sem: ind past} \\ \text{arg: } \textbf{[person z] [dog w]} \\ \ldots \\ \text{prn: 35} \end{bmatrix} \begin{bmatrix} \text{sur: fido} \\ \text{noun: } \textbf{[dog w]} \\ \text{cat: snp} \\ \text{sem: nm sg m} \\ \text{fnc: see} \\ \ldots \\ \text{prn: 35} \end{bmatrix}$$

The extrapropositional nature of the example is shown by the different prn values, from 33 to 35. The function word marking the slots is that:

6.4.2 MARKED SLOTS IN ITERATING OBJECT CLAUSES

The last clause terminates the iteration because the filler does not introduce another slot. The analyses 6.3.3, 6.3.4, and 6.4.1 rely on the strictly time-linear derivation order of DBS.

6.5 Marked Slot-Filler Repetition in Adnominal Clauses

Like infinitives and object clauses, adnominal clauses may be iterated, as in Mary saw the man who loves the woman who ... feeds Fido.

6.5.1 GRAPH ANALYSIS UNDERLYING MULTIPLE ADNOMINAL CLAUSES

Mary saw the man who loves the woman who fed Fido.

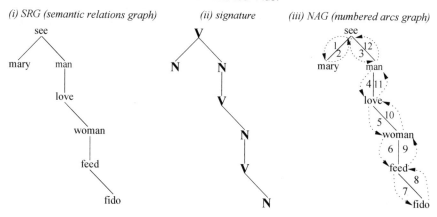

(i) SRG (semantic relations graph)   *(ii) signature*   *(iii) NAG (numbered arcs graph)*

(iv) surface realization (English, subject gap)

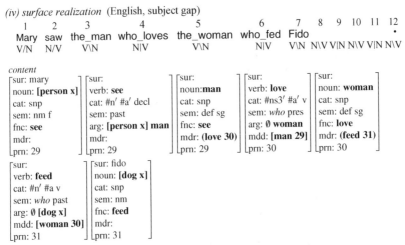

1	2	3	4	5	6	7	8	9	10	11	12
Mary	saw	the_man	who_loves	the_woman	who_fed	Fido					•
V/N	N/V	V\N	N\V	V\N	N\V	V\N	N\V	V\N	N\V	V\N	N\V

content

$$\begin{bmatrix} \text{sur: mary} \\ \text{noun: } \textbf{[person x]} \\ \text{cat: snp} \\ \text{sem: nm f} \\ \text{fnc: } \textbf{see} \\ \text{mdr:} \\ \text{prn: 29} \end{bmatrix} \begin{bmatrix} \text{sur:} \\ \text{verb: } \textbf{see} \\ \text{cat: } \#n' \ \#a' \ \text{decl} \\ \text{sem: past} \\ \text{arg: } \textbf{[person x] man} \\ \text{mdr:} \\ \text{prn: 29} \end{bmatrix} \begin{bmatrix} \text{sur:} \\ \text{noun:} \textbf{man} \\ \text{cat: snp} \\ \text{sem: def sg} \\ \text{fnc: } \textbf{see} \\ \text{mdr: } \textbf{(love 30)} \\ \text{prn: 29} \end{bmatrix} \begin{bmatrix} \text{sur:} \\ \text{verb: } \textbf{love} \\ \text{cat: } \#ns3' \ \#a' \ v \\ \text{sem: } who \text{ pres} \\ \text{arg: } \emptyset \ \textbf{woman} \\ \text{mdd: } \textbf{[man 29]} \\ \text{prn: 30} \end{bmatrix} \begin{bmatrix} \text{sur:} \\ \text{noun: } \textbf{woman} \\ \text{cat: snp} \\ \text{sem: def sg} \\ \text{fnc: } \textbf{love} \\ \text{mdr: } \textbf{(feed 31)} \\ \text{prn: 30} \end{bmatrix}$$

$$\begin{bmatrix} \text{sur:} \\ \text{verb: } \textbf{feed} \\ \text{cat: } \#n' \ \#a \ v \\ \text{sem: } who \text{ past} \\ \text{arg: } \emptyset \ \textbf{[dog x]} \\ \text{mdd: } \textbf{[woman 30]} \\ \text{prn: 31} \end{bmatrix} \begin{bmatrix} \text{sur: fido} \\ \text{noun: } \textbf{[dog x]} \\ \text{cat: snp} \\ \text{sem: nm} \\ \text{fnc: } \textbf{feed} \\ \text{mdr:} \\ \text{prn: 31} \end{bmatrix}$$

In this construction, the function word marking the slot is a "relative pronoun" (subordinating conjunction with argument role) such as who, whom, or which.[4] As shown by the increasing prn values, the construction is extrapropositional.

6.5.2 MARKED SLOT STRUCTURE OF ITERATING ADNOMINAL CLAUSES

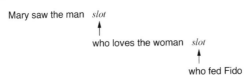

Mary saw the man *slot*

who loves the woman *slot*

who fed Fido

The function word marking the slots in this example is who.

6.6 Unmarked Slot-Filler Iteration in Gapping Constructions

In gapping constructions, a single unmarked slot is used several times, as in subject and predicate gapping (filler-slot), and object gapping (slot-filler). This is in contradistinction to slot-filler iterations which use several different marked slots (6.3–6.5).

In subject gapping, a single subject slot takes multiple predicate fillers, which may be shown as follows:

6.6.1 SUBJECT GAPPING AS AN ABSTRACT COGNITIVE STRUCTURE

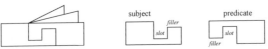

Without the multiple predicates, this three-dimensional graph would equal a simple subject-predicate combination, i.e., a subject with a single intransitive or transitive

---

[4] For more on "relative clauses" see TExer 3.3, 3.4.

verb. Computationally, the slot-filler combination is implemented as a cross-copying between order-free proplets.

Instead of marking the slots with a function word, as in slot-filler iteration, gapping marks the slots with a pause in speech or a comma in writing. This kind of slot is called 'unmarked,' but indicated by ∅ for analysis:

6.6.2 DBS ANALYSIS OF SUBJECT GAPPING (TExer 5.2)

Bob bought an apple, ∅ peeled a pear, and ∅ ate a peach.

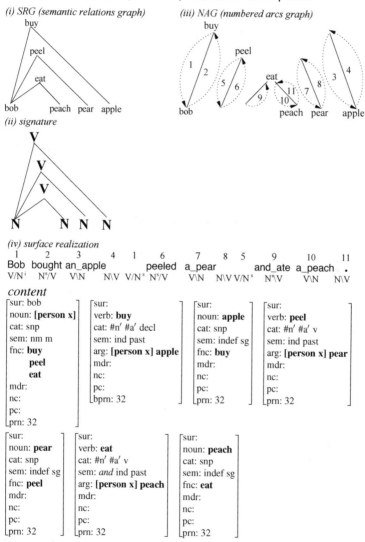

(i) SRG (semantic relations graph)

(ii) signature

(iii) NAG (numbered arcs graph)

(iv) surface realization

1	2	3	4	1	6	7	8	5	9	10	11
Bob	bought	an_apple			peeled	a_pear			and_ate	a_peach	.
V/N^i	N^a/V	V\N	N\V	V/N^s	N^a/V	V\N	N\V V/N^s		N^a\V	V\N	N\V

content

$$\begin{bmatrix} \text{sur: bob} \\ \text{noun: [person x]} \\ \text{cat: snp} \\ \text{sem: nm m} \\ \text{fnc: buy} \\ \quad\text{peel} \\ \quad\text{eat} \\ \text{mdr:} \\ \text{nc:} \\ \text{pc:} \\ \text{prn: 32} \end{bmatrix}$$
$$\begin{bmatrix} \text{sur:} \\ \text{verb: buy} \\ \text{cat: \#n' \#a' decl} \\ \text{sem: ind past} \\ \text{arg: [person x] apple} \\ \text{mdr:} \\ \text{nc:} \\ \text{pc:} \\ \text{bprn: 32} \end{bmatrix}$$
$$\begin{bmatrix} \text{sur:} \\ \text{noun: apple} \\ \text{cat: snp} \\ \text{sem: indef sg} \\ \text{fnc: buy} \\ \text{mdr:} \\ \text{nc:} \\ \text{pc:} \\ \text{prn: 32} \end{bmatrix}$$
$$\begin{bmatrix} \text{sur:} \\ \text{verb: peel} \\ \text{cat: \#n' \#a' v} \\ \text{sem: ind past} \\ \text{arg: [person x] pear} \\ \text{mdr:} \\ \text{nc:} \\ \text{pc:} \\ \text{prn: 32} \end{bmatrix}$$

$$\begin{bmatrix} \text{sur:} \\ \text{noun: pear} \\ \text{cat: snp} \\ \text{sem: indef sg} \\ \text{fnc: peel} \\ \text{mdr:} \\ \text{nc:} \\ \text{pc:} \\ \text{prn: 32} \end{bmatrix}$$
$$\begin{bmatrix} \text{sur:} \\ \text{verb: eat} \\ \text{cat: \#n' \#a' v} \\ \text{sem: and ind past} \\ \text{arg: [person x] peach} \\ \text{mdr:} \\ \text{nc:} \\ \text{pc:} \\ \text{prn: 32} \end{bmatrix}$$
$$\begin{bmatrix} \text{sur:} \\ \text{noun: peach} \\ \text{cat: snp} \\ \text{sem: indef sg} \\ \text{fnc: eat} \\ \text{mdr:} \\ \text{nc:} \\ \text{pc:} \\ \text{prn: 32} \end{bmatrix}$$

In contradistinction to the depth-first numbering of the NAGs in marked slot-filler constructions, the NAG numbering in gapping constructions is breadth-first (TExer

1). The final gapped item is announced by the function word *and*, coded in the sem slot of the *eat* proplet.

The proplet *bob* serves as the shared subject of the predicates buy, peel, and eat by specifying them in its fnc slot as the *gap list*. The inverse relation from the predicates to their shared subject is established by writing the core value [person x] of *bob* into the first (subject) arg slot of the verbs. The construction is intrapropositional, as indicated by the shared prn value 32.

The next gapping construction is predicate gapping, in which a single predicate takes multiple subjects and objects as fillers (three-dimensional):

6.6.3 PREDICATE GAPPING AS AN ABSTRACT THOUGHT STRUCTURE

Without the multiple subjects and objects, the graph would equal a simple subject/predicate\object combination.

The canonical DBS graph analysis of predicate gapping may be shown as follows:

6.6.4 DBS ANALYSIS OF PREDICATE GAPPING (TExer 5.3)

Bob **bought** an apple, Jim Ø a pear, and Bill Ø a peach.

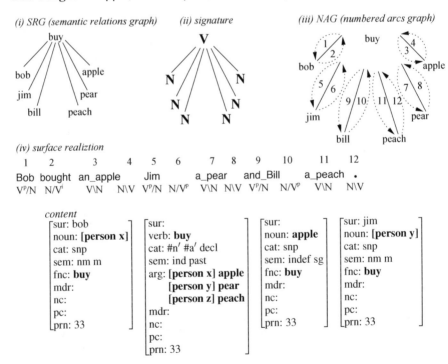

$$
\begin{bmatrix}
\text{sur:} \\
\text{noun: } \mathbf{pear} \\
\text{cat: snp} \\
\text{sem: indef sg} \\
\text{fnc: } \mathbf{buy} \\
\text{mdr:} \\
\text{nc:} \\
\text{pc:} \\
\text{prn: 33}
\end{bmatrix}
\begin{bmatrix}
\text{sur: bill} \\
\text{noun: } [\mathbf{person\ z}] \\
\text{cat: snp} \\
\text{sem: } and \text{ nm m} \\
\text{fnc: } \mathbf{buy} \\
\text{mdr:} \\
\text{nc:} \\
\text{pc:} \\
\text{prn: 33}
\end{bmatrix}
\begin{bmatrix}
\text{sur:} \\
\text{noun: } \mathbf{peach} \\
\text{cat: snp} \\
\text{sem: indef sg} \\
\text{fnc: } \mathbf{buy} \\
\text{mdr:} \\
\text{nc:} \\
\text{pc:} \\
\text{prn: 33}
\end{bmatrix}
$$

The shared item is the predicate *buy*. Its arg slot contains the gap list, here the subject-object pairs *bob apple, jim pear,* and *bill peach*. The conjunction *and* is coded into the initial sem slot of *bill*. The subject and object proplets take buy as their shared fnc value.

In object gapping a single object is taken by multiple predicates. Its graph is the mirror image of the three-dimensional graph for subject gapping (6.6.1):

6.6.5 OBJECT GAPPING AS AN ABSTRACT THOUGHT STRUCTURE

Without the multiple predicates, the graph would equal a simple object\predicate combination (including a predicate without a subject, as in imperatives).

6.6.6 DBS ANALYSIS OF OBJECT GAPPING (TExer 5.4)

Bob bought ∅, Jim peeled ∅, and Bill ate a peach

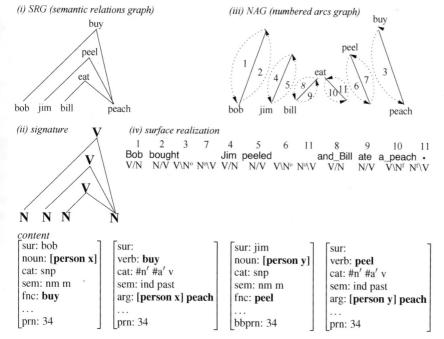

(i) SRG (semantic relations graph)

(iii) NAG (numbered arcs graph)

(ii) signature

(iv) surface realization

1	2	3	7	4	5	6	11	8	9	10	11
Bob	bought			Jim	peeled			and_Bill	ate	a_peach	•
V/N	N/V	V\N°	N°\V	V/N	N/V	V\N°	N°\V	V/N	N/V	V\Nf	Nf\V

content

$$
\begin{bmatrix}
\text{sur: bob} \\
\text{noun: } [\mathbf{person\ x}] \\
\text{cat: snp} \\
\text{sem: nm m} \\
\text{fnc: } \mathbf{buy} \\
\ldots \\
\text{prn: 34}
\end{bmatrix}
\begin{bmatrix}
\text{sur:} \\
\text{verb: } \mathbf{buy} \\
\text{cat: \#n' \#a' v} \\
\text{sem: ind past} \\
\text{arg: } [\mathbf{person\ x}]\ \mathbf{peach} \\
\ldots \\
\text{prn: 34}
\end{bmatrix}
\begin{bmatrix}
\text{sur: jim} \\
\text{noun: } [\mathbf{person\ y}] \\
\text{cat: snp} \\
\text{sem: nm m} \\
\text{fnc: } \mathbf{peel} \\
\ldots \\
\text{bbprn: 34}
\end{bmatrix}
\begin{bmatrix}
\text{sur:} \\
\text{verb: } \mathbf{peel} \\
\text{cat: \#n' \#a' v} \\
\text{sem: ind past} \\
\text{arg: } [\mathbf{person\ y}]\ \mathbf{peach} \\
\ldots \\
\text{prn: 34}
\end{bmatrix}
$$

$$
\begin{bmatrix}
\text{sur: bill} \\
\text{noun: } \textbf{[person z]} \\
\text{cat: snp} \\
\text{sem: } \textit{and} \text{ nm m} \\
\text{fnc: } \textbf{eat} \\
\text{mdr:} \\
\text{nc:} \\
\text{pc:} \\
\text{prn: 34}
\end{bmatrix}
\begin{bmatrix}
\text{sur:} \\
\text{verb: } \textbf{eat} \\
\text{cat: \#n}' \text{ \#a}' \text{ decl} \\
\text{sem: ind past} \\
\text{arg: } \textbf{[person z] peach} \\
\text{mdr:} \\
\text{nc:} \\
\text{pc:} \\
\text{prn: 34}
\end{bmatrix}
\begin{bmatrix}
\text{sur:} \\
\text{noun: } \textbf{peach} \\
\text{cat: snp} \\
\text{sem: indef sg} \\
\text{fnc: } \textbf{[p. x] buy} \\
\quad\quad\quad \textbf{[p. y] peel} \\
\quad\quad\quad \textbf{[p. z] eat} \\
\text{. . .} \\
\text{prn: 34}
\end{bmatrix}
$$

Gapping constructions are intrapropositional, like iterated infinitives.

6.7 Long-Distance Dependency

The slot-filler iteration of object clauses (6.4) may be combined with a single filler-slot relation, taking the filler in first and the slot in last position. This results in the following long-distance dependency (TExer 5.5):

6.7.1 OBJECT-CLAUSE ITERATION WITH LONG-DISTANCE DEPENDENCY

Whom did John say that Bill believes that Mary loves?

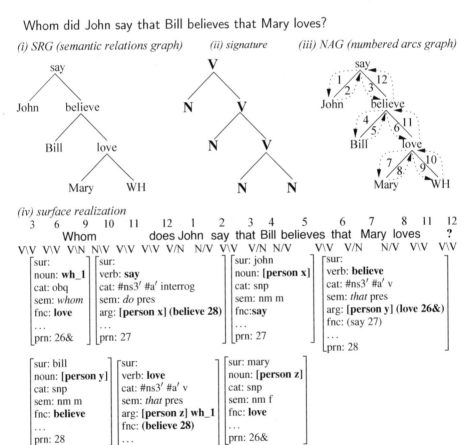

(i) SRG (semantic relations graph)   *(ii) signature*   *(iii) NAG (numbered arcs graph)*

(iv) surface realization

As in the object clause iteration 6.4.1, the slots are marked with the function word that.

In DBS, an unbounded dependency has the form of an unbounded suspension, here between initial whom and the final transitive verb form loves. Because the length of the suspension is only determined at the end of the input sentence, the time-linear surface-compositional hear mode derivation has the following syntactic ambiguity structure (NLC 7.6.5):

6.7.2 AMBIGUITY STRUCTURE OF AN UNBOUNDED SUSPENSION

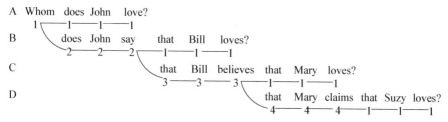

This systematic ambiguity does not affect the linear time complexity of DBS because the time-linear derivation replaces each previous reading with the new input. Accordingly, 6.7.2 has only a single reading, namely D.

6.8 Conclusion

This chapter investigates grammatical structures which are conspicuous in that they are based on repeated sharing. The question is whether these constructions are (i) limited to a certain language type or (ii) universal in that they may be found in numerous typologically unrelated natural languages. The latter hypothesis is supported by the fact that the iterations discussed have direct structural counterparts in several languages completely unrelated to the European languages, namely Korean[5], Tagalog[6], and Georgian[7]. This is remarkable, and analyzed data from additional languages supporting or opposing the conjecture would be interesting.

---

[5] Thanks to Professor Kiyong Lee, Ph.D., native speaker of Korean.
[6] Thanks to Mrs. Guerly Söllch, M.A., native speaker of Tagalog.
[7] Thanks to Mrs. Sofia Tekmalaze-Fornwald, M.A., native speaker of Georgian.

7. Computational Pragmatics

In agent-based data-driven DBS, propositions do not denote truth values. Instead they are content built (i) from the classical semantic kinds *concept, indexical*, and *name*, connected (ii) by the classical semantic relations of *subject/predicate, object\predicate, adnominal|noun, adverbial|verb*, and *conjunct−conjunct*, (iii) coded by address, (iv) at the elementary, phrasal, and clausal level of grammatical complexity,

Computational pragmatics[1] relies on the on-board orientation system, which is part of the agent's interface component and monitors moment by moment as a sequence of STARs. A STAR is a flat feature structure with the attributes S for space (location), T for time, A for agent (speaker), R for recipient (hearer), 3rd for pro3, and prn for proposition number.

A STAR is connected to a content type by a shared prn value, resulting in a content token. A content type with a language-dependent surface is a *literal meaning$_1$*. A corresponding token used for inter-agent language communication is an *utterance meaning$_2$*.

There is literal (7.3) and nonliteral pragmatics (7.4, 7.5). Literal pragmatics is an obligatory change of perspective. Examples are the change from I am thirsty to I was thirsty in the STAR-0 STAR-1 transition of the speak mode and the change from I see you to you see me in the STAR-1 STAR-2 transition of the hear mode.

Nonliteral pragmatics is optional non-literal use. It applies the same inference deductively in the speak mode (input matching antecedent) and abductively in the hear mode (input matching consequent). Examples are syntactic mood adaptations, e.g., Could you pass the salt? for Pass the salt!, and metaphor, e.g., melt for disappear.

7.1 Four Kinds of Content in DBS

In DBS, a content is defined as a set (order-free) of proplets. As the computational data structure, proplets are defined as non-recursive feature structures with ordered attributes. The proplets of a proposition are connected by (i) semantic relations of structure coded by address and (ii) a shared prn value.

There are four kinds of content in DBS, called [−surface −STAR], [−surface +STAR], [+surface −STAR], and [+surface +STAR], illustrated as follows:

---

[1] Overviews of noncomputational pragmatics are Kempson (2001) and Horn&Ward eds. (2004).

© The Author(s), under exclusive license to Springer Nature Switzerland AG 2023
R. Hausser, *Ontology of Communication*, https://doi.org/10.1007/978-3-031-22739-4_7

7.1.1 NONLANGUAGE CONTENT TYPE: [−surface, −STAR]

$$
\begin{bmatrix} \text{sur:} \\ \text{noun: dog} \\ \text{cat: snp} \\ \text{sem: def sg} \\ \text{fnc: find} \\ \text{mdr:} \\ \text{nc:} \\ \text{pc:} \\ \text{prn: K} \end{bmatrix}
\begin{bmatrix} \text{sur:} \\ \text{verb: find} \\ \text{cat: \#n}'\text{ \#a}'\text{ decl} \\ \text{sem: past ind} \\ \text{arg: dog bone} \\ \text{mdr:} \\ \text{nc:} \\ \text{pc:} \\ \text{prn: K} \end{bmatrix}
\begin{bmatrix} \text{sur:} \\ \text{noun: bone} \\ \text{cat: snp} \\ \text{sem: indef sg} \\ \text{fnc: find} \\ \text{mdr:} \\ \text{nc:} \\ \text{pc:} \\ \text{prn: K} \end{bmatrix}
$$

This proposition is a type because there is no STAR and the prn value is a variable, here K. It is a nonlanguage content because the sur slots are empty.

The next example is a corresponding nonlanguage token:

7.1.2 NONLANGUAGE CONTENT TOKEN: [−surface, +STAR]

$$
\begin{bmatrix} \text{sur:} \\ \text{noun: dog} \\ \text{cat: snp} \\ \text{sem: def sg} \\ \text{fnc: find} \\ \text{mdr:} \\ \text{nc:} \\ \text{pc:} \\ \text{prn: 12} \end{bmatrix}
\begin{bmatrix} \text{sur:} \\ \text{verb: find} \\ \text{cat: \#n}'\text{ \#a}'\text{ decl} \\ \text{sem: past ind} \\ \text{arg: dog bone} \\ \text{mdr:} \\ \text{nc:} \\ \text{pc:} \\ \text{prn: 12} \end{bmatrix}
\begin{bmatrix} \text{sur:} \\ \text{noun: bone} \\ \text{cat: snp} \\ \text{sem: indef sg} \\ \text{fnc: find} \\ \text{mdr:} \\ \text{nc:} \\ \text{pc:} \\ \text{prn: 12} \end{bmatrix}
\begin{bmatrix} \text{S: yard} \\ \text{T: friday} \\ \text{A: sylvester} \\ \text{R:} \\ \text{3rd:} \\ \text{prn: 12} \end{bmatrix}
$$

The three content proplets and the STAR proplet are connected by a common prn constant, here 12. According to the STAR, the token resulted as an observation by the agent Sylvester on Friday in the yard.

The language content type corresponding to 7.1.1 illustrates the independence of language-dependent sur values, here German, from the relatively language-independent placeholders (English base forms for convenience):

7.1.3 LANGUAGE CONTENT TYPE: [+SURFACE, −STAR]

$$
\begin{bmatrix} \text{sur: der\_Hund} \\ \text{noun: dog} \\ \text{cat: snp} \\ \text{sem: def sg} \\ \text{fnc: find} \\ \text{mdr:} \\ \text{nc:} \\ \text{pc:} \\ \text{prn: K} \end{bmatrix}
\begin{bmatrix} \text{sur: fand} \\ \text{verb: find} \\ \text{cat: \#n}'\text{ \#a}'\text{ decl} \\ \text{sem: past ind} \\ \text{arg: dog bone} \\ \text{mdr:} \\ \text{nc:} \\ \text{pc:} \\ \text{prn: K} \end{bmatrix}
\begin{bmatrix} \text{sur: einen\_Knochen} \\ \text{noun: bone} \\ \text{cat: snp} \\ \text{sem: indef sg} \\ \text{fnc: find} \\ \text{mdr:} \\ \text{nc:} \\ \text{pc:} \\ \text{prn: K} \end{bmatrix}
$$

A language content type is also called a literal meaning$_1$. It is an abstraction in that an actual DBS hear mode derivation results in a content token. However, a content type may always be obtained from a content token by removing the STAR and replacing the prn constants with suitable variables.

The fourth kind of content is a language token which matches the type, here 7.1.3, called an utterance meaning$_2$. Our example is produced by the speaker Sylvester in

German towards the intended hearer Tweety and corresponds to the nonlanguage content token 7.1.2 except for the R value:

7.1.4 LANGUAGE CONTENT TOKEN: [+surface, +STAR]

$$
\begin{bmatrix}
\text{sur: der\_Hund} \\
\text{noun: dog} \\
\text{cat: snp} \\
\text{sem: def sg} \\
\text{fnc: find} \\
\text{mdr:} \\
\text{nc:} \\
\text{pc:} \\
\text{prn: 12}
\end{bmatrix}
\begin{bmatrix}
\text{sur: fand} \\
\text{verb: find} \\
\text{cat: \#n' \#a' decl} \\
\text{sem: past ind} \\
\text{arg: dog bone} \\
\text{mdr:} \\
\text{nc:} \\
\text{pc:} \\
\text{prn: 12}
\end{bmatrix}
\begin{bmatrix}
\text{sur: einen\_Knochen} \\
\text{noun: bone} \\
\text{cat: snp} \\
\text{sem: indef sg} \\
\text{fnc: find} \\
\text{mdr:} \\
\text{nc:} \\
\text{pc:} \\
\text{prn: 12}
\end{bmatrix}
\begin{bmatrix}
\text{S: yard} \\
\text{T: friday} \\
\text{A: sylvester} \\
\text{R: tweety} \\
\text{3rd:} \\
\text{prn: 12}
\end{bmatrix}
$$

According to the STAR, the transfer of content occurred on Friday in the yard. The content types 7.1.1 and 7.1.3 match not only the tokens 7.1.2 and 7.1.4, but an open number of corresponding tokens with different prn values.[2]

An utterance meaning$_2$ exists in the cognition of the speaker, and – if transfer is successful – of the hearer. The raw data serving as the vehicle of transfer in communication, in contrast, have absolutely no meaning or grammatical properties whatsoever at all (no reification in DBS), but may be measured by natural science.

7.2 Coactivation Resulting in Resonating Content

The basis of literal and nonliteral pragmatic interpretation is the automatic coactivation of *resonating content* in the agent's on-board database. Resonating content is the computational counterpart of association in psychology and relies on (a) the database schema of the agent's A-memory and (b) the data structure of proplets.

7.2.1 TWO-DIMENSIONAL DATABASE SCHEMA OF A-MEMORY IN DBS

- *horizontal*
  Proplets with the same core value are stored in the same token line in the time-linear order of their arrival.
- *vertical*
  Token lines are in the alphabetical order induced by their core value's letter sequence.

---

[2] The type/token distinction applies not only to propositions, but also to the sign kinds. A concept like square is a type if the length value is a variable, and a token if it is a constant. An indexical like you is a type if a STAR proplet is absent in the proposition and otherwise a token. A name like Lucy is a type if the named referent value is absent and otherwise a token.

[2] The STAR in agent-based DBS may be seen as a development of the sign-based "parameter approach," which uses infinite sets such as I (possible worlds), J (possible moments of time), S (possible speakers), H (possible hearers), etc. as parameters. In Montague (1973) the index @, I, J, g is superscripted to logical formulas (possible world semantics). Cresswell (1972, p.4) wonders tongue in cheek about adding a next drink parameter.

In the agent's A-memory, the time-linear arrival order of proplets is reflected by the position in their token line and by their prn value. The sequence of (i) *member proplets* is followed by a free slot as part of the column called the (ii) *now front*, and the (iii) *owner*.[3]

7.2.2 SCHEMATIC EXAMPLE OF TOKEN LINE WITH CLEARED NOW FRONT

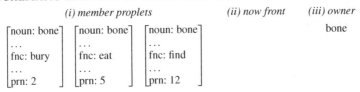

Consider the content 7.1.2 as stored at the now front before clearance:

7.2.3 STORAGE OF A PROPOSITION AT THE NOW FRONT

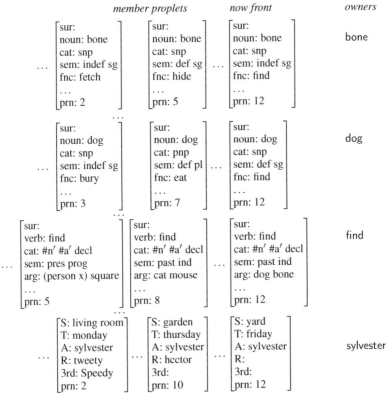

The storage of a proplet at the now front uses the letter sequence of the core value for accessing the correct token line via the owner. Retrieval searches the token line using

---

[3] The terminology of member proplets and owner values is reminiscent of the member and owner records in a classic network database (Elmasri and Navathe (2017)), which inspired the content-addressable database schema of the DBS A-memory (formerly called word bank).

the prn value. Content is automatically activated by computing successor proplets based on continuation addresses.

When the proplets at the current now front have ceased to be candidates for additional processing (e.g., cross-copying), the now front is cleared by moving it and the owners one step to the right into fresh memory territory (loom-like clearance). By leaving the proplets of the completed proposition, here with the prn value 12, behind, their location becomes permanent storage as member proplets, like sediment, never to be changed. The only way to correct is adding content, like a diary entry. The corrected version of a content is the most recent one, i.e., rightmost, in the token lines involved.

In summary, the A-memory illustrated in 7.2.3 is content-addressable because it does not use a separate index (catalog), unlike a coordinate-addressable database, e.g., an RDBMS. The key for a proplet's storage in and retrieval from the agent's memory is not a location, but the letter sequence of the proplet's core value (which enables computational string search in combination with a trie structure). Based on reference by address, A-memory automatically monitors the changes in a content.

7.3 Literal Pragmatics of Adjusting Perspective

In agent-based monitoring, a current content recedes inevitably into the past. Therefore there are two STARs, the STAR of origin attached to a content now past, called the STAR-0, and the current STAR of the agent's looking back at the past content in memory, called the STAR-1. For example, if the STAR-0 content I see you is retrieved from memory it must be changed to the STAR-1 content I saw you. Also, the values Paris and Thursday of the STAR-0, for example, may change to London and Sunday of the STAR-1, and similarly for the R and the 3rd value.

In language communication, the speaker uses the STAR-1 to encode the past STAR-0 content into a language surface. The hearer's interpretation of the speaker's STAR-1 surface, however, necessitates a third STAR, called the STAR-2. For example, the STAR-1 content I saw you must be changed to the STAR-2 content You saw me. The (i) origin, (ii) production, and (iii) interpretation of a content used in language communication are in an obligatory order called the temporal backbone:

7.3.1 TEMPORAL BACKBONE OF DBS

agent A		*agent A*		*agent B*
	conversion–1		*conversion–2*	
STAR–0		STAR–1		STAR–2
				▶ *time line*
origin		production		interpretation
I see you.		I saw you.		You saw me.

The agent's STAR-0 values Paris and Friday, for example, may change into the speak mode's STAR-1 values London and Sunday, which in turn may change to New York and Wednesday of the hearer's STAR-2, and similarly for the R and the 3rd value.

To illustrate the changes from a STAR-0 content of origin to a STAR-1 content of production, let us begin with the following example:

7.3.2 STAR-0 CONTENT OF ORIGIN: I see you.

STAR-0 proplet of origin

$$
\begin{bmatrix} \text{sur:} \\ \text{noun: pro1} \\ \text{cat: s1} \\ \text{sem: sg} \\ \text{fnc: see} \\ \text{mdr:} \\ \text{nc:} \\ \text{pc:} \\ \text{prn: 12} \end{bmatrix}
\begin{bmatrix} \text{sur:} \\ \text{verb: see} \\ \text{cat: \#n-s3}' \text{ \#a}' \text{ decl} \\ \text{sem: pres ind} \\ \text{arg: pro1 pro2} \\ \text{mdr:} \\ \text{nc:} \\ \text{pc:} \\ \text{prn: 12} \end{bmatrix}
\begin{bmatrix} \text{sur:} \\ \text{noun: pro2} \\ \text{cat: sp2} \\ \text{sem:} \\ \text{fnc: see} \\ \text{mdr:} \\ \text{nc:} \\ \text{pc:} \\ \text{prn: 12} \end{bmatrix}
\begin{bmatrix} \text{S: yard} \\ \text{T: thursday}^4 \\ \text{A: sylvester} \\ \text{R: hector} \\ \text{3rd:} \\ \text{prn: 12} \end{bmatrix}
$$

By definition, a STAR-0 content is (i) without language (quasi language-independent) and (ii) the verb's sem value is pres ind. The pro1 and the pro2 indexicals in the content point at the A value, here sylvester, and the R value, here hector, respectively, of the STAR-0 proplet.

In the speak mode, the agent's STAR-0 content I see you (7.3.2) may be mapped automatically into one of the following STAR-1 variants. They differ semantically, but are pragmatically equivalent (CC 7):

7.3.3 PRAGMATICALLY EQUIVALENT STAR-1 CONTENTS

STAR-1 a: Sylvester remembers the content 7.3.2 without speaking.
STAR-1 b: Sylvester tells Hector that he saw him.
STAR-1 c: Sylvester tells Speedy that he saw "him," referring to Hector.
STAR-1 d: Sylvester tells Speedy that he saw Hector.
STAR-1 e: Sylvester tells Speedy that he saw Hector in the yard.
STAR-1 f: Sylvester tells Speedy that he saw Hector on Thursday in the yard.

In variants a-d, the semantic differences in the contents are compensated pragmatically by varying STAR-1 values. In e and f, past STAR-0 values overwritten by the current ones are preserved by writing them into the content (CC 7.1).

In the hear mode, each STAR-1 variant in 7.3.3 must be mapped into an equivalent STAR-2 content (CC 8):

7.3.4 PRAGMATICALLY EQUIVALENT STAR-2 CONTENTS

STAR-2 a: <For nonlanguage content, a hear mode counterpart does not exist>
STAR-2 b: Hector understands that Sylvester saw him.
STAR-2 c: Speedy understands that Sylvester saw "him," i.e., Hector.
STAR-2 d: Speedy understands that Sylvester saw Hector.
STAR-2 e: Speedy understands that Sylvester saw Hector on Thursday.
STAR-2 f: Speedy understands that Sylvester saw Hector on T. in the yard.

The speaker's STAR-0 STAR-1 conversion-1 and the hearer's STAR-1 STAR-2 conversion-2 use different inferences for the interpretation of pronominal indexicals. The reason is that the interpretation of pronominals in the speaker's conversion-1 is unchanged, but must be inverted in the hearer's conversion-2.

The pragmatic equivalence of the semantically different contents 7.3.2, 7.3.3, and 7.3.4 is based (i) on the choice between coding certain values either in the content as concepts (including the named referent of names), or as (ii) indexicals pointing at the S, T, R, and 3rd values of the STAR. Space and Time information may be coded solely in the STAR, as in variants a-d in 7.3.3 and b-d in 7.3.4, or also be written explicitly into the content, as in the variants e and f.

7.4 Nonliteral Pragmatics of Syntactic Mood Adaptation

The literal pragmatic inferences described in the previous section are obligatory insofar as the STAR-0, STAR-1, and STAR-2 contents must be pragmatically equivalent in order for communication to succeed. Thereby the STAR-0 STAR-1 transitions (speak mode) and the STAR-1 STAR-2 transitions (hear mode) each require their own inference.

We turn now to two kinds of nonliteral pragmatics, called syntactic mood adaptation and figurative use. In contradistinction to literal pragmatics, (i) they are optional and (ii) their STAR-0 STAR-1 (speak mode) and STAR-1 STAR-2 (hear mode) transitions use the same inference, but inductively in the speak mode and abductively in the hear mode. In order for communication to succeed, the hearer must revert the speaker's nonliteral content back into the speaker's original content modulo the obligatory hear mode adjustments of a STAR-2 content.

Consider the following example from J.L. Austin ([1955]1962) of a syntactic mood adaptation:

7.4.1 SYNTACTIC MOOD ADAPTATION IMP-INT

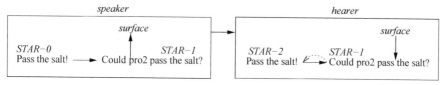

The speaker's communicative purpose of the STAR-0 STAR-1 conversion is softening a command (imperative[5]) into a polite request (yes-no interrogative). If the hearer

---

[4] Using weekdays as T values may be crude as compared to nano- or pico-seconds, but is sufficient for the current purpose.

[5] The subject of imperatives in English and many other languages is implicit: it is automatically assumed to be pro2 in the speak mode and pro1 in the hear mode, without any surface manifestation. Consequently, the hearer's standard STAR-2 reversal from pro2 to pro1 is implicit as well.

were to take the speaker's STAR-1 content literally by answering yes or no, communication would fail. For communication to succeed, the hearer must use the same inference as the speaker, but abductively ($\nearrow$).

Syntactic mood adaptions are common in the European languages, but not universal.[6] The following two examples are based on the inferences INT-DECL (CC 7.4.2) and IMP-DECL (CC 7.5.2), respectively.

7.4.2 SYNTACTIC MOOD ADAPTATION INT-DECL

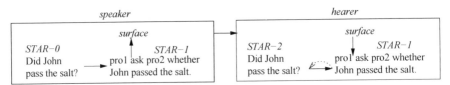

In this and the next example, the speaker's purpose is emphasis.

7.4.3 SYNTACTIC MOOD ADAPTATION IMP-DECL

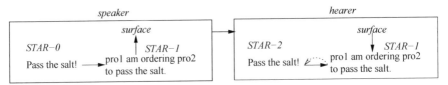

There are more syntactic-mood adaptations for the purpose of emphasis, such as the interrogative-imperative adaptation from Did you pass the salt? to Tell me if you passed the salt![7]

7.5 Nonliteral Pragmatics of Figurative Use

While the inferences of syntactic mood adaptation use complete propositions as input and output, the inferences of figurative use only a part, elementary or phrasal, of a proposition. Figurative use is subject to the invariance constraint (CC 6.4.4), according to which the figurative replacement must be of the same syntactic category and the same semantic field as the literal original. The condition that successful communication requires the hearer to reconstruct the speaker's original content applies equally to syntactic mood adaptation and figurative use.

In figurative use, lexical relations such as hyponymy, metonymy, property sharing, abbreviation, and membership in the same semantic field serve as the basis of inferencing (CC 9). Consider the following examples:

---

[6] The alternative in Korean, for example, is the use of two morphological systems, one for honor and one for mood, which are agglutinated to the verbal stem. Thanks to Prof. Kiyong Le for his help in this matter.

7.5.1 FIGURATIVE USE BASED ON HYPONYMY RELATION

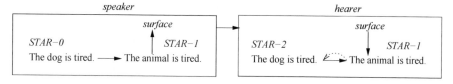

7.5.2 FIGURATIVE USE BASED ON SHARED PROPERTY INFERENCE

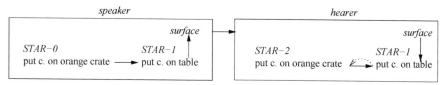

7.5.3 SPEAK AND HEAR MODE OF AN ABBREVIATING ADVERBIAL USE

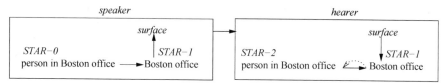

7.5.4 SPEAK AND HEAR MODE OF AN ABBREVIATING ADNOMINAL USE

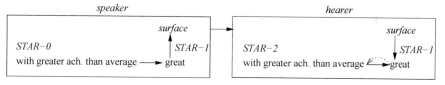

As in all abductive use, there is no certainty regarding the output of the inference (i.e., of the antecedent). For example, if the orange crate in 7.5.2 were accompanied by a footstool, a sideboard, and a low bookshelf, it would be impossible for the hearer to decide what the speaker meant with "table" (too many candidates with flat horizontal surfaces in the current context of interpretation, embarrassment of riches). In such a case, the speaker would have to specify more precisely what is meant in order for communication to succeed.

7.6 Conclusion

In agent-based data-driven DBS, the semantic/pragmatic distinction is based on the type/token distinction from philosophy. The semantics of a content is a type which is

---

[7] The dependence of nonliteral use pragmatics on specific pronouns, concepts, tense, mood, etc. is reminiscent of Construction Grammar (Fillmore 1988).

independent of the utterance situation, while the pragmatics is a token which connects a type to the agent's on-board orientation system (OBOS).

There are two kinds of pragmatics in DBS. The *obligatory literal* kind adapts a semantic content to the alternative perspectives of speaker and hearer. This requires different inferences for the speak and the hear mode (7.3).

The other kind is *optional nonliteral* pragmatics. It provides informative views such as metaphor on literal content. In successful communication, the hearer must reconstruct the speaker's literal content from which the nonliteral view was derived. For this, speaker and hearer use the same inference, though deductively in the speak mode and abductively in the hear mode (7.4).

8. Discontinuous Structures in DBS and PSG

In linguistics, grammatical constructions with semantically connected word forms are called 'discontinuous' if the word forms are not adjacent. For example, in (i) Peter looked the number up, the semantically related 'looked...up' are separated by 'the_number', and in (ii) Yesterday Mary danced the semantically related 'Yesterday...danced' are separated by 'Mary'.

This chapter explores why example (i) poses a descriptive problem for PSG (Phrase Structure Grammar), but not for DBS (Database Semantics), and why example (ii) poses a descriptive problem for DBS, but not for PSG. Then the PSG and DBS proposals for solving their respective problem are compared.

8.1 The Time-Linear Structure of Natural Language

Many syntactic-semantic content structures in human cognition are hierarchical, but the transfer of content in language communication is strictly linear. In the medium of speech, this holds for direct language communication, such as talking face to face, and for indirect language communication, such as talking on the phone:

8.1.1 COMPARING FACE-TO-FACE WITH ON THE PHONE TALKING

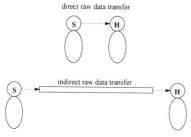

The cognitive aspects are located inside the agents' heads, with **S** for speaker, **H** for hearer, and the arrow heads indicating the direction of transfer. Communication operates the same regardless of whether the agents are face to face (direct raw data transfer) or talking on the phone (indirect raw data transfer).

All that is required of an artificial or natural transfer channel is the transmission of data without distortion (Shannon and Weaver 1948). However, though the transfer channel is not the place for reconstructing cognition (pace Eco[1] 1975), it poses a crucial structural requirement for language communication: the signs must be in a linear order (canonized by de Saussure ([1916]1972) as his *second principe*). This is because humans can neither produce nor interpret two or more words, phrases, sentences, or texts simultaneously.

Cognitive structures may be as hierarchical as needed as long as language content can be coded in the time-linear order fit for the transfer channel.[2] The speak mode takes a content as input and produces a surface sequence as output by traversing the input content along the address-coded semantic relations between proplets. The hear mode takes a surface sequence as input and produces an output content by concatenating proplets by address into a set. Because the speak and the hear mode take different kinds of input and output, they can never use the same algorithm.

Consider the hear mode taking the unanalyzed *surface* The dog snored. as input for the time-linear surface-compositional derivation producing the output *content*:

8.1.2 DERIVING CONTENT FROM SURFACE IN HEAR MODE

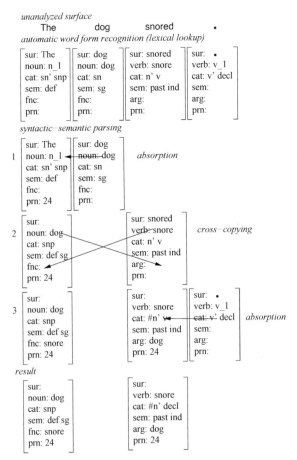

Automatic word form recognition of the hearer's interface component provides a sequence of four lexical proplets, unconnected with nonempty sur slots. The derivation provides the output, i.e., a content of two nonlexical proplets, connected and with empty sur slots. The derivation order is (i) *time-linear*[3](left-associative), shown by the stair-like addition of next word proplets, the analysis is (ii) *surface compositional* because each lexical item has a concrete sur value and there are no surfaces without a proplet analysis, and the activation and application of operations is (iii) *data-driven* by the lexical input proplets.

Defining a content as a set of proplets connected by address is essential for storage in and retrieval from the agent's on-board memory (content-addressable database). The following example illustrates the order-free coding of the subject/predicate relation:

8.1.3 ORDER-FREE PROPLETS CODING THE CONTENT OF The dog snored.

$$
\begin{bmatrix}
\text{sur:} \\
\text{noun: } \mathbf{dog} \\
\text{cat: def sg} \\
\text{sem:} \\
\text{fnc: } \mathbf{snore} \\
\text{mdr:} \\
\text{nc:} \\
\text{pc:} \\
\text{prn: 24}
\end{bmatrix}
\qquad
\begin{bmatrix}
\text{sur:} \\
\text{verb: } \mathbf{snore} \\
\text{cat: \#n' decl} \\
\text{sem: past ind} \\
\text{arg: } \mathbf{dog} \\
\text{mdr:} \\
\text{nc:} \\
\text{pc:} \\
\text{prn:24}
\end{bmatrix}
$$

Computationally, semantic relations of structure are established by cross-copying addresses. Intuitively, semantic relations are shown graphically in four views:

8.1.4 GRAPHICAL PRESENTATION OF THE SEMANTIC RELATIONS IN 8.1.2

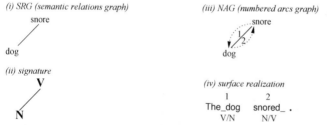

The *(i) SRG* and the *(ii) signature* on the left show the static aspect of the hierarchical content structure. The *(iii) NAG (numbered arcs graph)* and *(iv) surface realization* on the right show its dynamic traversal.

---

[1] Eco's most basic prototype of communication is a buoy "telling" an engineer the water level of a lake; the engineer's partner of discourse is mother nature. Grice's (1957) "bus bell model" and Dretsky's (1981) "doorbell model" have another human ringing the bell (CLaTR 2.2.4).

[2] This requirement is not fulfilled by formal grammars which compute possible substitutions, such as PSG and CG. For this reason, Chomsky emphasizes tirelessly (e.g., Chomsky 1965, p. 9) that Generative Grammar (nativism) models the innate structure of natural language (structuralism), and is "not intended" for a transfer of content in language communication ("autonomy of syntax", Chomsky 1982). It is somewhat unlikely, however, that the innate structure of natural language would do without a speak mode, and a hear mode, and a transfer channel, especially in language acquisition, as shown by the analogy with anatomy.

[3] Aho and Ullman (1977), p. 47; FoCL 10.1.1.

The arc numbers of the *NAG* are used for specifying (1) a think mode navigation and (2) a think-speak mode surface production as shown by the *(iv) surface realization*. The traversal operations in the think mode of DBS are (1) predicate╱subject, (2) subject╱predicate, (3) predicate╲object, (4) object╲predicate, (5) noun↓adnominal, (6) adnominal↑noun, (7) verb↓adverbial, (8) adverbial↑verb, (9) noun→noun, (10) noun←noun, (11) adnominal→adnominal, (12) adnominal←adnominal, (13) adverbial→adverbial, (14) adverbial←adverbial, and (15) verb→verb..

The *(iv) surface realization* shows language-dependent production. It is implemented as the speak mode riding piggy-back on the think mode navigation. The agent's memory provides the concepts of dog and snore as types (*declarative specification*). The agent's interface component adapts the surface types into tokens and realizes them as raw data (*operational implementation*).

8.2 Constituent Structure Paradox of PSG

DBS and PSG differ ontologically in that DBS is agent-based data-driven, while PSG is sign-based substitution-driven. The derivations of PSG generate different language expressions from the same node, called S for sentence or start. The number of possible tree structures for a given surface grows exponentially with the length of the surface. Assuming binary branching, there are two PSG trees for a three-word sentence:

8.2.1 DIFFERENT PSG TREES FOR THREE-WORD UNAMBIGUOUS SURFACE

From a formal point of view, both trees are equally well-formed.

However, because more than one tree for an unambiguous surface does not make sense linguistically, there must be an intuitive principle for choosing the "good" one. In chomskyan linguistics, this is the principle of Constituent Structure[4]:

8.2.2 DEFINITION[5] OF CONSTITUENT STRUCTURE

1. Words or constituents which belong together semantically must be dominated directly and exhaustively by a node.

2. The lines of a Constituent Structure may not cross (*nontangling condition*).

---

[4] Compared to Aristotle, Constituent Structure is quite recent. It evolved from the *immediate constituent analysis* of L. Bloomfield (1887–1949). His student Z. Harris (1909–1992) turned constituent analyses into substitution and movement tests. Harris's student N. Chomsky turned the methodologically motivated tests into generative rules, called transformations and proffered as innate.

[5] Provided by Prof. Ivan Sag, personal communication, Stanford 1989–91.

Assuming that knows and John (predicate-object) belong closer together semantically than Julia and knows (subject-predicate), the tree on the left satisfies 8.2.2 and is therefore considered linguistically correct, while the tree on the right is not. Yet it has been known at least since 1953 (Bar-Hillel 1964, p. 102) that there are certain natural language constructions with "discontinuous elements" which violate the definition of Constituent Structure.[6]

Known as the Constituent Structure Paradox (FoCL 8.5), the problem may be illustrated with discontinuous look_ _up in the following attempts to analyze Peter looked the number up as a Constituent Structure:

8.2.3 VIOLATING THE SECOND CONDITION OF 8.2.1

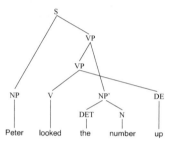

Here the semantically related expressions looked and up are dominated directly and exhaustively by a node, satisfying the first condition of 8.2.1. The analysis violates the second condition, however, because the lines cross.[7]

The alternative attempt satisfies the second condition, but violates the first:

8.2.4 VIOLATING THE FIRST CONDITION OF 8.2.1

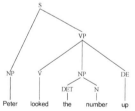

Here the lines of the tree do not cross, satisfying the second condition, but the semantically related expressions looked − up, or rather the nodes V and DE dominating them, are not *exhaustively* dominated by a node. Instead, the node directly dominating V and DE also dominates the NP the_number.

---

[6] Bloomfield's (1933, p.210) analysis of gentle/man/ly argues for constituent structure. Along the same lines, we could take Latin te video (I see you) and use the morphological (inflectional) coding of the subject role and the syntactic coding of the object role as an argument against the Structuralist's general assumption of what belongs semantically more closely together.

[7] DPSG (Discontinuous Phrase Structure Grammar, Bunt et al. 1987) argues for accepting crossing lines in Phrase Structure Trees (quasi three-dimensional). This was preceded by pleas for using only context-free phrase structure by Harman 1963, McCawley 1982a, Gazdar et al. (1985), and others.

In summary, the output of the context-free base in transformational grammar must satisfy Constituent Structure but presumably[8] can not accommodate some natural surfaces. The output of the transformation component must satisfy the natural surfaces, but sometimes violates Constituent Structure. Worst of all, the introduction of transformations raises the computational complexity degree from polynomial (n^3) of context-free PSG to undecidable (Peters&Ritchie 1973).

In DBS, the example shown in 8.2.3 and 8.2.4 for PSG derives without problem in the time-linear, surface-compositional, data-driven manner of the hear mode:

8.2.5 DISCONTINUOUS STRUCTURE DERIVATION IN DBS HEAR MODE

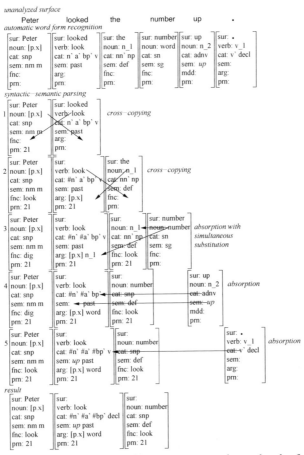

The semantic relations in the resulting content are shown by the following graph analysis. The surface realization of the speak mode is in English:

8.2.6 GRAPHICAL PRESENTATION OF THE SEMANTIC RELATIONS IN 8.2.5

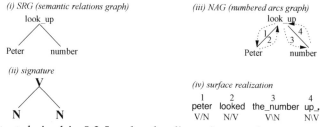

(i) SRG (semantic relations graph)

(ii) signature

(iii) NAG (numbered arcs graph)

(iv) surface realization

The content derived in 8.2.5 codes the discontinuous element *up* as the initial sem value of the *look* proplet in line 4. In the *(iv) surface realization*, up_ • is realized from the finite verb (goal proplet of arc 4 in the *(iii) NAG*).

8.3 Suspension in Database Semantics

The reason why Peter looked the number up violates Constituent Structure but Yesterday Mary danced does not may be shown by comparing their respective context-free PSG derivations, based on the possible substitutions of rewrite rules:

8.3.1 CONTEXT-FREE PSGS FOR TWO DISCONTINUOUS STRUCTURES

Peter looked the number up

S	→ NP VP
VP	→ VP NP
VP	→ V NP DE
NP	→ N
NP	→ DET N
N	→ Peter
V	→ looked
DET	→ the
N	→ number
DE	→ up

Yesterday Mary danced

S	→ ADV VP
VP	→ NP V
NP	→ N
ADV	→ yesterday
N	→ Mary
V	→ danced

The left PSG for 8.2.4 violates the requirement of *exhaustive* dominance with the rule VP→V NP DE. The PSG on the right, in contrast, generates Yesterday Mary danced without any problem: first S→ADV VP places the ADV in initial position; then VP→NP V places the subject noun after the ADV and before the V.

The apparent problem of DBS with Yesterday Mary danced, in contrast, results from the time-linear derivation order. It creates a temporary situation in which the modifier yesterday can not be connected because the modified danced has not yet arrived. The solution is a *suspension* until danced becomes available:

---

[8] Perhaps surprisingly, there exists a context-free PSG analysis for Peter looked the number up which satisfies 8.2.2. It is shown in 8.5.1 on the left, opposite the tree for Yesterday Mary danced.

8.3.2 HEAR MODE DERIVATION OF Yesterday Mary danced.

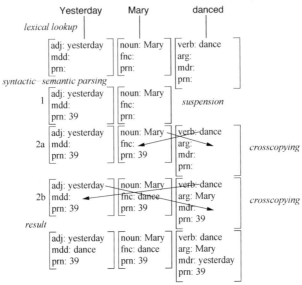

Because operations are data-driven in DBS, instances of suspension are compensated automatically, without any need for additional software (convergence in science).

The phenomenon is asymmetric because the semantically equivalent word order Mary danced+yesterday requires neither suspension nor absorption (adjacency).

8.3.3 HEAR MODE DERIVATION OF Mary danced yesterday.

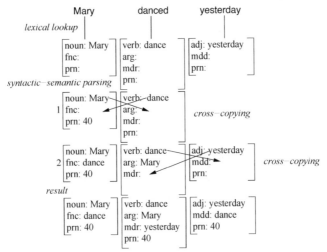

Suspension at the elementary level scales up directly to the phrasal and clausal levels. At the clausal level, the relation is extrapropositional:

8.3.4 INTERPRETATION OF When Fido barked Mary laughed

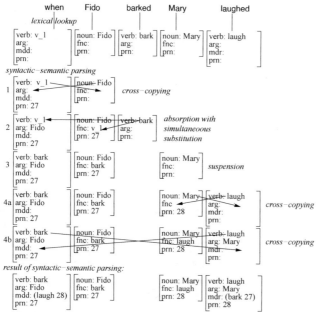

The suspension occurs in line 3 and is compensated in 4a and 4b.

If the optional modifier clause follows, there is no suspension, just as in 8.3.3:

8.3.5 INTERPRETATION OF Mary laughed when Fido barked

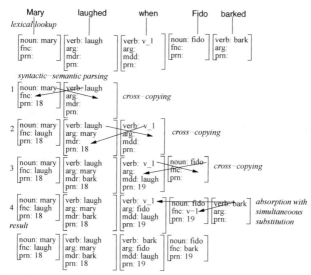

Consider the semantic relations graph which underlies both surfaces:

8.3.6 GRAPHICAL PRESENTATION OF SEMANTIC RELATIONS IN 8.3.4 AND 8.3.5

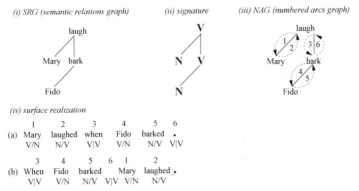

(i) SRG (semantic relations graph)   *(ii) signature*   *(iii) NAG (numbered arcs graph)*

(iv) surface realization

	1	2	3	4	5	6
(a)	Mary	laughed	when	Fido	barked	.
	V/N	N/V	V|V	V/N	N/V	V|V

	3	4	5	6	1	2
(b)	When	Fido	barked	Mary	laughed	.
	V|V	V/N	N/V	V|V	V/N	N/V

The suspension at the elementary level 8.3.4 and its variant 8.3.5 at the clausal level of grammatical complexity show that there are natural surface orders in which suspension cannot be avoided. Regarding the linear complexity of LAG/DBS (TCS'82), ambiguities induced by suspension are benign because they are not recursive.

It seems that discontinuous filler-slot constructions require a suspension 8.4.1, whereas discontinuous slot-filler constructions make do with an absorption (i.e., without ambiguity, 8.4.2). An example is extrapropositional coordination, as in a text. The relation between two adjacent sentences holds between the top verb of sentence n (into which the interpunctuation has been absorbed) and the top verb of sentence n+1. In a sequence of English declaratives, for example, the verb of sentence n+1 is usually preceded by the subject or an adverbial, creating the discontinuity.

The pivot of the absorption transition from the verb of sentence n to the verb of sentence n+1 is the interpunctuation (function word). This is shown by the following hear mode derivation of Julia sleeps. John sings. Suzy dreams.:

8.3.7 ABSORPTIONS IN EXTRAPROPOSITIONAL COORDINATION

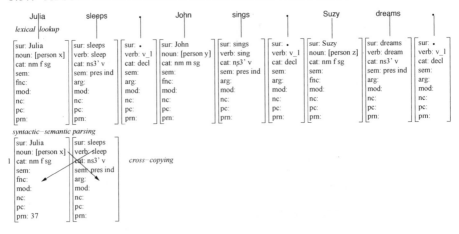

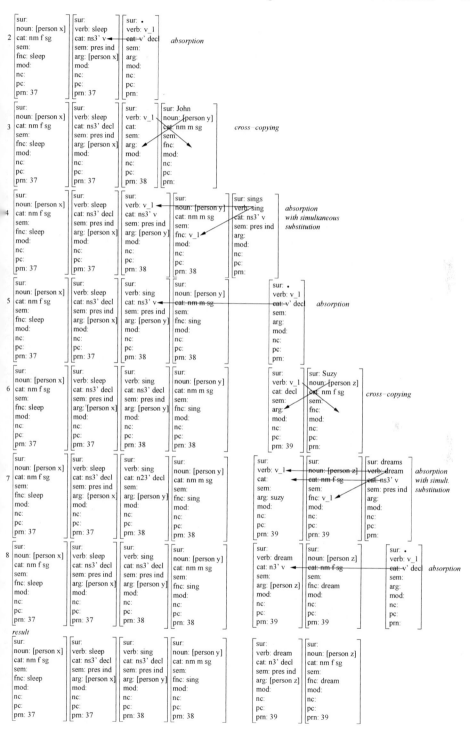

Despite two extrapropositional transitions with intervening subjects (lines 3 and 6), there are no suspensions and consequently no compensations. Instead, the full stop proplets in lines 2 and 5 simultaneously (i) supply the verbal mood decl to the verb of proposition n, i.e., the one the full-stop belongs to, and (ii) cross-copy with the subject of proposition n+1 if there is one. However, if there is no subsequent proposition, a variant of the interpunct operation (TExer 2.1.5–2.1.19) discards the interpunctuation proplet once its verbal mood value has been utilized (line 8).

8.4 Discontinuity With and Without Suspension in DBS

The domain-range structure of a semantic relation may be viewed as a filler-slot constellation, with the domain providing the filler (also called the argument, actant, or complement) and the range (also called functor, codomain, or valency[9] carrier) providing the slot. In (i) subject/predicate, the subject is the filler and the predicate provides the slot, in (ii) object\predicate, the predicate provides the slot and the object is the filler, in (iii) modifier|modified, the modifier is the filler and the modified provides the slot, and in (iv) coordination, the slot-filler relation between conjuncts is bidirectional.[10]

If filler and slot provider are adjacent in the hear mode, i.e., if there are no intervening items, order does not make a difference. However, if there are intervening items and the slot provider precedes, then (a) the slot defines the kind of compatible filler, (b) the relation is initiated, and (c) the hearer can simply wait until the filler arrives and plops into place. If the filler precedes, in contrast, there may be many kinds of slot providers which is why no relation can be initiated; instead the hearer must wait until automatic word form recognition provides the slot, finally enabling an on the spot filler-slot combination.

In English filler-slot constellations, the intervening items must be bridged by a suspension. Consider the following examples:

8.4.1 SUSPENSION IN 'FILLER PRECEDES SLOT' DISCONTINUITIES

1. Clausal subject precedes main clause
   That Fido found a bone surprised Mary.[11]
2. Adverbial precedes predicate (8.3.2)
   Yesterday Mary danced.
3. Clausal modifier precedes main clause (8.3.4)
   When Fido barked Mary laughed.

---

[9] L. Tesnière (1959).

[10] The explicit specification of the semantic relations of structure by means of the connectives /, \, |, and − is more concise than the "belong semantically together" intuition of nativism.

[11] In That Fido barked amused Mary (TExer, 2.5), the V/V relation between *bark* (intransitive verb) and *amuse* is not discontinuous because there is no grammatical object (no suspension needed). It is similar in the V|A relation of Mary danced yesterday (8.3.3).

4. Subject gapping (TExer, 5.2)
   Bob bought an apple, peeled a pear, and ate a peach.
5. Predicate gapping (TExer, 5.3)
   Bob bought an apple, Jim a pair, and Bill a peach.
6. Long distance dependency (TExer, 5.5)
   Who(m) did John say that Bill believes that Mary loves?

The grammatical role of the function words that, when, who, and and is shown in detail in the explicit hear mode derivations referred to.

Discontinuous 'slot provider precedes filler' constructions, in contrast, do not require a suspension. Instead, an *absorption* suffices:

8.4.2 ABSORPTION IN 'SLOT PRECEDES FILLER' DISCONTINUITIES

1. Verb precedes bare preposition (TExer, 4.3)
   Fido dug the bone up.
2. Main clause precedes clausal object (TExer, 2.6)
   Mary heard that Fido barked.
3. Main clause precedes clausal modifier (8.3.5)
   Mary laughed when Fido barked.
4. Object gapping (TExer, 5.4)
   Bill bought, Jim peeled, and Bill ate a peach.
5. Repeating object clauses (TExer, 5.6)
   Mary saw the man who loves the woman who fed Fido.
6. Period precedes subject in extrapropositional coordination (8.3.7):
   Julia sleeps. John sings. Suzy dreams.

In gapping constructions, however, the relation between the order of filler and slot provider, and the use of suspension vs. absorption is the reverse as compared to 8.4.1 and 8.4.2:

8.4.3 ABSORPTION VS. SUSPENSION IN GAPPING

1. Subject gapping (TExer, 5.2)
   Bob bought an apple, peeled a pear, and ate a peach.
   Exception: filler-slot without suspension
2. Predicate gapping (TExer, 5.3)
   Bob bought an apple, Jim a pair, and Bill a peach.
   Exception: filler-slot without suspension
3. Object gapping (TExer, 5.4)
   Bill bought, Jim peeled, and Bill ate a peach.
   Exception: slot-filler with suspension

Object gapping is special in that it requires not only suspension, but also a derivation-external cache for storing the gap providers until the filler (as the standard location of the gap list) arrives.

8.5 Conclusion

This chapter started with an enigma. The examples
> (i) Peter *looked* the number *up.* and
> (ii) *Yesterday* Maria *danced.*

are both discontinuous. For context-free PSG (i) poses a problem, solved by adding a movement component, while (ii) runs straight through. For DBS, (i) runs straight through, while (ii) poses a problem, solved by suspension and data-driven cross-copying (8.3.2). Let us conclude by showing what applying the PSG solution for example (ii) (8.3.1) to example (i) would look like:

8.5.1 PHRASE STRUCTURE OF EXAMPLE (I) IN CONCORD WITH 8.2.2

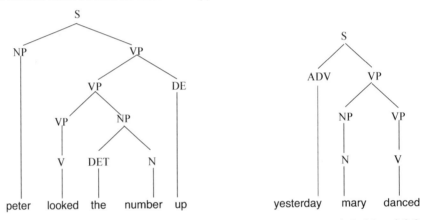

If the phrase structure on the right satisfies the Constituent Structure definition 8.2.2 then so does the one on the left: in both trees the lines do not cross; also the rules (a) VP → VP DE used on the left and (b) S → ADV VP used on the right are alike in that two nodes which "belong together semantically" are dominated directly and exhaustively by a single node.[12]

Today's PSG interprets "belonging together semantically" as the functor-argument relation of symbolic logic (first order predicate calculus). Thereby the question of whether the functor precedes or follows in the surface is treated as the language-dependent "problem of linearization." In data-driven DBS, in contrast, the functor-argument relation introduces a surface-compositional asymmetry, namely between a *slot-filler* and a *filler-slot* relation.

---

[12] It is perhaps not unlikely that the linguists working with constituent structure at the time were secretly aware of the obvious possibility shown by the analysis on the left of 8.5.1. What is so wrong with this phrase structure that adding a movement component was preferred even at the price of making the grammar algorithm undecidable? First, being undecidable was fashionable in substitution-driven complexity analysis of the time, e.g., Post's (1946) Correspondence Problem. Second, discontinuities were not the main motivation for introducing a movement component; instead it was the assumption of a universal, innate context-free base in combination with a transformation component (nativism). Thus a discontinuity analysis exceeding the generative power of context-free PSGs came in handy as "empirical" support for the Standard Theory (ST, Chomsky 1965).

A slot provider looking for a filler is like trying a bag of old keys on a given piece of furniture, while a filler looking for a slot is like trying a given key on an open number of furniture pieces. In English declaratives, the potential inefficiency of the filler-slot order in the subject/predicate constellation is effectively avoided by the post-nominative position (FoCL 18.3) of the finite verb, which provides adjacency.[13]

---

[13] Thanks to Professor Kiyong Lee for helpful comments.

9. Classical Syllogisms as Computational Inferences

The classical categorical syllogisms originated and evolved in times without computers, sensors for vision and audition, and actuators for manipulation and vocalization.
Therefore, a sign-based substitution-driven ontology was the only practical option.

That it is nevertheless possible to translate categorical syllogisms into the agent-
based data-driven inferencing of DBS is because they use the same set-theoretic structures. The following reconstruction proceeds from the diagrams by Swiss mathemati-
cian Leonard Euler, used in the year 1761 in a famous 'letter to a princess.'

9.1 Logical vs. Common Sense Reasoning

A basic distinction in analytic philosophy is between logical reasoning and common
sense reasoning. Logical reasoning is based on set theory, which is why the associated
inferences in DBS are called S-inferences. Common sense reasoning, in contrast, is
without a set-theoretic aspect, and the associated inferences are called C-inferences.

In the human prototype, S-inferences and C-inferences are not separated, but work
smoothly together. Therefore, the computational model of reasoning in DBS uses the
same general inference schema and the same data structure for S- and C-inferences.
Consider the following comparison of the two kinds of DBS inferences as schematic
examples:

9.1.1 EXAMPLE OF AN S-INFERENCE (FERIO)

S-inference: α is homework $\Rightarrow$ α is no fun
                        $\Uparrow$ $\Downarrow$
            input: some reading is homework   output: some reading is no fun

9.1.2 EXAMPLE OF A C-INFERENCE (CAUSE_AND_EFFECT)

C-inference: α is hungry $\Rightarrow$ α is cranky
                      $\Uparrow$ $\Downarrow$
            input: Laura is hungry   output: Laura is cranky

Both inferences work by binding the subject term of the input to the variable α in the
antecedent of the pattern, which enables the consequent to derive the output.

R. Hausser, *Ontology of Communication*, https://doi.org/10.1007/978-3-031-22739-4_9

The implementation of the syllogism FERIO as an S-inference is illustrated in 9.6.10, while the corresponding implementation of the C-inference 9.1.2 is shown as the following software operation:

9.1.3 APPLYING THE C-INFERENCE 9.1.2 IN DBS FORMAT

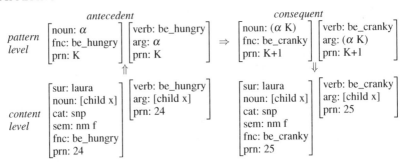

The content and the pattern level consist of nonrecursive feature structures with ordered attributes. Called proplets[1], they serve as the computational data structure. The proplets of a content are order-free but connected by the classical semantic relations of structure, i.e., functor-argument and coordination, coded by address.

The proplets at the pattern level use variables as values for the 'core' and the 'continuation' attributes, those at the content level have corresponding constants. For computational pattern matching to be successful (i) the attributes of the pattern proplet must be a *sublist*, (ii) the variables of the pattern proplet must be *compatible*, and (iii) the constants of the pattern proplet must be *identical* with those of the corresponding content proplet directly underneath.

S-inferences and C-inferences differ in the source of their reasoning. For example, in the S-inference 9.1.1, the source is the disjunction between the concepts homework and be_fun and the intersection between reading and homework (9.6.8), which are assumed to be generally accepted. In the C-inference 9.1.2, in contrast, the source is something observed by the agent(s).[2]

9.2 Categorical Syllogisms

An early highlight in the Western tradition of logical reasoning are the classical syllogisms of Aristotle (384–322 BC) and their further development by the medieval scholastics.[3] In the modern era, the syllogisms have been based on the intuitions of set theory (Euler, 9.2.3).

---

[1] So-called because they are the elementary items of propositions.

[2] The resulting set-theoretic relation between being cranky and being hungry in 9.1.2, i.e., intersection, is merely a consequence of the reasoning, and not the source.

[3] For a critical review of how the understanding of Aristotle's theory of categorical syllogisms changed over the millenia see Read (2017). For a computational automata and factor analysis see Zhang Yinsheng and Qiao Xiaodong (2009).

A categorical[4] syllogism consists of three parts, called premise 1, premise 2, and the conclusion. This may be shown schematically as follows:

9.2.1 SCHEMATIC INSTANTIATION OF A CATEGORICAL SYLLOGISM

> Major premise: all M are P
> Minor premise: all S are M
> Conclusion: all S are P

M is the middle term, S the subject, and P the predicate. M is shared by the two premises. The positions of M are called the alignment of a syllogism.

The three parts of a classical syllogism are restricted to the four categorical judgments, named **A, E, I,** and **O** by the Scholastics:

9.2.2 THE FOUR CATEGORICAL JUDGMENTS

A	universal affirmative	$\forall x\, [\, f(x) \rightarrow g(x)\,]$	all f are g
E	universal negative	$\neg \exists x\, [\, f(x) \wedge g(x)\,]$	no f are g
I	particular affirmative	$\exists x\, [\, f(x) \wedge g(x)\,]$	some f are g
O	particular negative	$\exists x\, [\, f(x) \wedge \neg g(x)\,]$	some f are not g

The first-order Predicate Calculus representation in the third column is in a linear notation called *prenex normal form*, which superseded Frege's (1879) graphical format.

The four categorical judgments combine systematically into 256 (2^8) possible syllogism, of which 24 have been found valid. The syllogisms reconstructed in this chapter as DBS inferences are BARBARA,[5] CELARENT, DARII, FERIO, BAROCO, and BOCARDO, plus the modi ponendo ponens and tollendo tollens as special[6] cases.

The set-theoretic constellations underlying the four categorical judgments may be shown as follows:

---

[4] The term *categorical* refers to the strict specification of the Aristotelian syllogisms, especially in their medieval form, such as exactly two premises – one conclusion, middle term not in the conclusion, subject/predicate structure of the three parts, using only the four categorical judgments, etc.

[5] The scholastics used the vowels of the categorical judgments in the names of the associated syllogisms as mnemonic support. For example, the three vowels in the name of modus BARBARA indicate that the categorical judgments of the propositions serving as the two premises and the consequent are all of the kind **A**, i.e., universal affirmative (9.2.2).

[6] The modi ponendo ponens (9.3.1) and tollendo tollens (9.4.1) are not categorical syllogism in the narrow sense because their premise 2 and conclusion are not categorical judgments of the kind **A, E, I,** or **O**. This is reflected by their different naming convention as compared to categorical syllogisms proper, for example BARBARA or FERIO.

9.2.3 Set-theoretic counterparts of categorical judgments

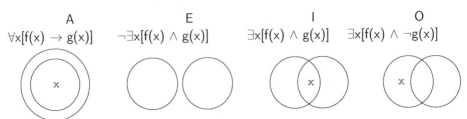

Known as Euler diagrams[7] (Euler 1761), the set-theoretic constellations are used in DBS to reconstruct the valid syllogisms as agent-based data-driven S-inferences. As an example, consider the schematic application of modus Barbara in DBS:

9.2.4 Modus Barbara as a DBS inference

A DBS inference takes a given content as input and validly derives a new content as output. Validity follows from set-theoretic intuitions which are the foundation of both the sign-based classical syllogisms and their agent-based DBS counterparts.

The transition from a categorical syllogism to a DBS inference may be shown as follows:

9.2.5 From syllogism to DBS inference

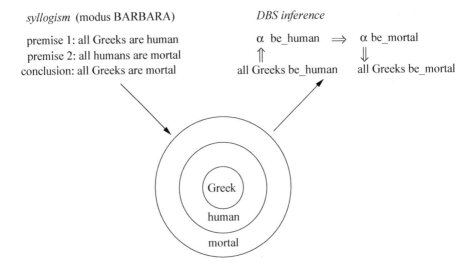

The DBS reconstruction of a categorical syllogism as an inference has the form $\alpha\ X$ *implies* $\alpha\ Y$, whereby the variable α represents a grammatical subject, and X and

Y represent grammatical predicates. The α in the antecedent may be matched by and bound to (1) a complete set, e.g., all Greeks (universal, 9.4.4), (2) a subset, e.g., some pets (particular, 9.6.4), or (3) an element, e.g., Socrates (individual, 9.6.3). In the consequent, the input-binding of α derives the output.

With the possible presence of negation in the antecedent, the consequent, or both, there result the following four schemata of S-inferences for the categorical syllogisms, each with a universal, a particular, and an individual variant.

The first triple is without negation:

9.2.6 α BE_X IMPLIES α BE_Y

(1) universal: all Greeks be_human implies all Greeks be_mortal.
(2) particular: some pets be_rabbits implies some pets be_furry.
(3) individual: Socrates be_human implies Socrates be_mortal.

The universal version is modeled after modus BARBARA (9.5.1), the particular version after modus DARII (9.6.1), and the individual version after modus ponendo ponens (9.3.1).

The second triple negates the consequent:

9.2.7 α BE_X IMPLIES α NOT BE_Y

(4) universal: all horses be_quadruped implies all horses not be_human.
(5) particular: some pets be_turtles implies some pets not be_furry.
(6) individual: Pegasus be_quadruped implies Pegasus not be_human.

The universal version is modeled after modus CELARENT (9.5.6), the particular version after modus FERIO (9.6.6), and the individual version after modus tollendo tollens (9.4.1).

The third triple negates the antecedent:

9.2.8 α NOT BE_X IMPLIES α BE_Y

(7) universal: all friars not be_married implies all friars be_single.
(8) particular: some men not be_married implies some men be_single.
(9) individual: Fred not be_married implies Fred be_single.

Set-theoretically, the denotations of not be_ married and of be_ single are coextensive in all three versions.

The fourth triple negates the antecedent and the consequent. Though EEE syllogisms are not valid for all instantiations, the following instantiations are:

---

[7] Named after Leonhard Euler (1707–1783). The method was known already in the 17th century and has been credited to several candidates.

　　Venn (1881, p.113) called Euler diagrams "old-fashioned". Euler diagrams reflect the set-theoretic constellations simple and direct, whereas Venn models the complicated medieval superstructures erected by the scholastics on top of the original syllogisms. Venn diagrams are useful for showing that certain syllogisms, for example, EEE-1 and OOO-1, are not valid.

9.2.9 α NOT BE_X IMPLIES α NOT BE_Y

(10) universal: all gods not be_mortal implies all gods not be_human.
(11) particular: some pets not be_furry implies some pets not be_rabbits.
(12) individual: Zeus not be_mortal implies Zeus not be_human.

Set-theoretically, the denotations of not be_X and not be_Y are disjunct in the (10) universal and the (12) individual variant, and in the complement of the pet-rabbit intersection in the (11) particular variant.

9.3 Modus Ponendo Ponens

Modus ponendo[8] ponens serves as the individual version of 9.2.6. The standard representation in Predicate Calculus is as follows:

9.3.1 MODUS PONENDO PONENS IN PREDICATE CALCULUS

> premise 1: $\forall x[f(x) \rightarrow g(x)]$
> premise 2: $\exists y[f(y)]$
> ─────────────────
> conclusion: $\exists z[g(z)]$

Instantiating f as be_human and g as be_mortal has the following result:

9.3.2 INSTANTIATING MODUS PONENDO PONENS

> premise 1: For all x, if x is human, then x is mortal.
> premise 2: There exists a y, such that y is human.
> ─────────────────
> conclusion: There exists a z, such that z is mortal.

The reconstruction of modus ponendo ponens (NLC 5.3) in DBS (i) turns premise 1 into the form α is human implies α is mortal, called the inference, (ii) uses premise 2 as the input, and (iii) treats the conclusion as the output:

9.3.3 REPHRASING MODUS PONENDO PONENS IN DBS

> inference: α be_human implies α be_ mortal.
> input: Socrates be_human.[9]
> output: Socrates be_mortal.

Shown here with input for modus ponens (individual), the inference works just as well for particular (9.2.6, 1) and universal (9.2.6, 2) input.

---

[8] In Propositional Calculus, modus ponendo ponens and modus tollendo ponens differ as follows: modus ponendo ponens has the premises (A→B) and A, resulting in the conclusion B; modus tollendo ponens has the premises (A∨B) and ¬A, also resulting in B. The distinction disappears in the functor-argument structure of DBS.

Using the DBS data structure, the inference applies as follows to the modus ponendo ponens input of 9.3.3:

9.3.4 APPLYING MODUS PONENDO PONENS AS FORMALIZED IN DBS

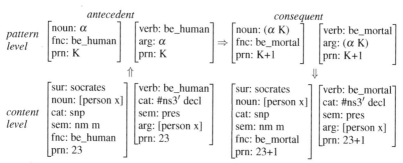

The DBS reinterpretation of premise 1 as the inference and premise 2 as the input requires that the input be compatible for matching with the antecedent. This would be prevented, however, if the antecedent specified the noun pattern α as a plural, corresponding to $\forall x$ in premise 1 of 9.3.1, and premise 2 as a singular, corresponding to $\exists y$. Therefore, the noun pattern α in 9.3.4 omits the cat and sem features, thus enabling matching (compatibility by omission). By vertically binding the constant socrates of the content level to the variable α in the antecedent of the pattern level, the consequent derives the new content socrates is mortal as output.

9.4 Modus Tollendo Tollens

Modus tollendo[9] tollens serves as the individual version of 9.2.7. A standard representation in Predicate Calculus is as follows:

9.4.1 MODUS TOLLENDO TOLLENS IN PREDICATE CALCULUS

> premise 1: $\forall x[f(x) \rightarrow g(x)]$
> premise 2: $\exists y[\neg g(y)]$
> conclusion: $\exists z[\neg f(z)]$

Let us instantiate f as is human and g as is biped:

---

[8] Predicate Calculus treats the copula-adnominal combination is human as the elementary proposition be_human(x) which denotes a truth value. DBS, in contrast, analyzes is human as the modifier|modified (or rather modified|modifier) combination is|human (CC 4.6). For comparison, simplicity, and brevity we compromise here by using the Predicate Calculus notation, e.g., be_human, like intransitive verbs as values in proplets, but without any variable.

[9] In Propositional Calculus, modus ponendo ponens and modus tollendo ponens differ as follows: modus ponendo ponens has the premises $(A{\rightarrow}B)$ and A, resulting in the conclusion B; modus tollendo ponens has the premises $(A{\vee}B)$ and $\neg A$, also resulting in B. The distinction disappears in the functor-argument structure of DBS.

9.4.2 INSTANTIATING MODUS TOLLENDO TOLLENS

premise 1: For all x, if x is human, then x is biped.
premise 2: there exists a y which is not biped.
conclusion: There exists a z which is not human.

If we use quadruped to instantiate not be_biped and horse to instantiate not be_human, the set-theoretic constellation underlying modus tollendo tollens in 9.4.2 may be depicted as follows:

9.4.3 SET-THEORETIC VIEW OF MODUS TOLLENDO TOLLENS

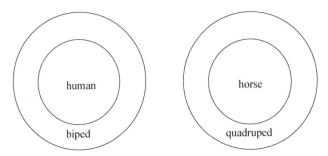

Premise 1 corresponds to the set structure on the left, premise 2 to the set structure on the right (with pegasus as the individual instantiation of horse, and the sets horse and human being disjunct).

The DBS reconstruction uses the disjunction of the sets quadruped and human for the inference α be_quadruped implies α not be_human. Pegasus being an element of the set quadruped renders the input Pegasus is quadruped. Pegasus not being an element of the set human renders the output Pegasus not be_human.

9.4.4 REPHRASING MODUS TOLLENDO TOLLENS IN DBS

inference: α be_quadruped implies α not be_human.
    input: Pegasus be_quadruped.
    output: Pegasus not be_human.

The inference works for the individual, the particular, and the universal variant of 9.2.7. The variants differ solely in their input and output.

Let us conclude with the translation of 9.4.4 into the data structure of DBS:

9.4.5 APPLYING MODUS TOLLENDO TOLLENS AS FORMALIZED IN DBS

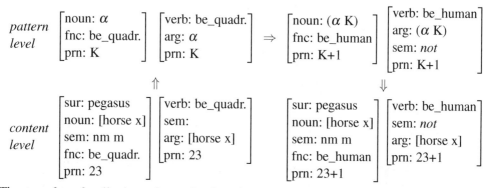

The transfer of syllogisms from sign-based substitution-driven symbolic logic to agent-based data-driven DBS relies on DBS inferencing being part of the think mode, which may run detached from the agent's interface component (CC: mediated reference 3.1.3, sequential application 3.6.2).

9.5 Modi BARBARA and CELARENT

The vowels in the name of modus BARBARA indicate the categorical judgments of the propositions serving as the premises and the consequent, which are all of type **A**, i.e., universal affirmative (9.2.2).

9.5.1 MODUS BARBARA IN PREDICATE CALCULUS

> premise 1: $\forall x[f(x) \rightarrow g(x)]$
> premise 2: $\forall y[g(y) \rightarrow h(y)]$
> conclusion: $\forall z[f(z) \rightarrow h(z)]$

The middle term is g. If f is realized as be_Greek, g as be_human, and h as be_mortal, then the syllogism reads as follows:

9.5.2 INSTANTIATING MODUS BARBARA

> premise 1: For all x, if x be_Greek, then x be_human.
> premise 2: For all y, if y be_human, then y be_mortal.
> conclusion: For all z, if z are Greek, then z be_mortal.

The set-theoretic constellation underlying modus BARBARA in 9.5.2 may be depicted as follows:

9.5.3 SET-THEORETIC VIEW OF MODUS BARBARA

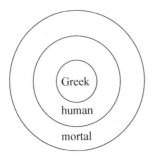

Premise 1 is expressed by the set Greek being a subset of human and premise 2 by the set Greek being a subset of mortal.

The DBS inference schema formulates the set-theoretic constellation as follows:

9.5.4 REPHRASING MODUS BARBARA IN DBS

> inference: α be_human implies α be_mortal.
>   input: All Greeks be_human.
>   output: All Greeks be_mortal.

The validity of the inference follows directly from the subset relations Greek $\subset$ human $\subset$ mortal, which are inherent in the extensions of these concepts.

Consider the translation of 9.5.4 into the data structure of DBS:

9.5.5 APPLYING MODUS BARBARA AS FORMALIZED IN DBS

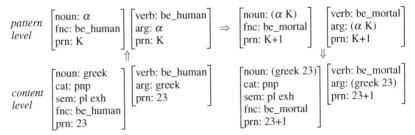

By vertically binding greek of the content level to the variable α in the antecedent of the pattern level, the consequent derives the desired new content All Greeks are mortal as output. In the class of syllogisms with unnegated antecedent and unnegated consequent (9.2.6), the reconstruction of BARBARA constitutes the universal, of DARII 9.6.5 the particular, and of modus ponendo ponens 9.3.4 the individual variant.

Next let us turn to a syllogism with the vowel **E** in its name, where **E** indicates a universal negative (9.2.2). The vowels in the name of modus CELARENT, for example, indicate that premise 1 is of type **E**, premise 2 of type **A**, and the conclusion of type **E**. In Predicate Calculus, CELARENT is represented as follows:

9.5.6 MODUS CELARENT IN PREDICATE CALCULUS

premise 1: $\neg\exists x[f(x) \wedge g(x)]$
premise 2:　$\forall y[h(y) \rightarrow f(y)]$
conclusion: $\neg\exists z[h(z) \wedge g(z)]$

The middle term is f. If f is realized as human, g as quadruped, and h as Greek, then
the syllogism reads as follows:

9.5.7 INSTANTIATING CELARENT IN PREDICATE CALCULUS

premise 1: $\neg\exists x[human(x) \wedge quadruped(x)]$
premise 2:　$\forall y[Greek(y) \rightarrow human(y)]$
conclusion: $\neg\exists z[Greek(z) \wedge quadruped(z)]$

The set-theoretic constellation underlying modus CELARENT in 9.5.7 may be depicted as follows:

9.5.8 SET-THEORETIC VIEW OF MODUS CELARENT

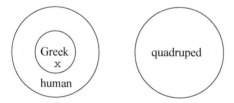

Premise 1 is expressed by the sets human and quadruped being disjunct. Premise 2 is
expressed by the set Greek being a subset of human. The conclusion is expressed by
the sets Greek and quadruped being disjunct.
    The inference schema of DBS describes the set-theoretic constellation as follows:

9.5.9 REPHRASING MODUS CELARENT IN DBS

inference: α be_human implies α not be_quadruped
    input: All Greeks be_human
    output: All Greeks not be_quadruped

Consider the translation of 9.5.9 into the data structure of DBS:

9.5.10 APPLYING MODUS CELARENT AS FORMALIZED IN DBS

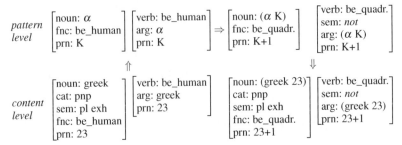

By binding greek of the content level to the variable α in the antecedent of the pattern level, the consequent derives the desired new content All Greeks are not quadruped as output. The $\forall x$ quantifier of Predicate Calculus is coded in the *greek* proplet by the feature [sem: pl exh] and the negation in the conclusion is coded in the predicate *be_quadruped* by the feature [sem: *not*].

9.6 Modi DARII and FERIO

The DBS variants of modus BARBARA (9.5.5) and modus CELARENT (9.5.10) have shown the treatment of the categorical judgments (9.2.2) **A** (universal affirmative) and **E** (universal negative). To show the treatment of the remaining categorical judgments **I** (particular affirmative) and **O** (particular negative), let us reconstruct the modi DARII and FERIO as DBS inferences.

The vowels in the name DARII indicate the categorical judgment **A** in premise 1, and **I** in premise 2 and the conclusion. The representation in Predicate Calculus is as follows:

9.6.1 MODUS DARII IN PREDICATE CALCULUS

> premise 1: $\forall x[f(x) \rightarrow g(x)]$
> premise 2: $\exists y[h(y) \wedge f(y)]$
> _____
> conclusion: $\exists z[h(z) \wedge g(z)]$

The middle term is f. If f is instantiated as be_rabbit, g as be_furry, and h as be_pet, then the syllogism reads as follows:

9.6.2 INSTANTIATING MODUS DARII

> premise 1: For all x, if x is rabbit, then x is furry.
> premise 2: For some y, y is pet and y is rabbit.
> _____
> conclusion: For some z, z is pet and z is furry.

The set-theoretic constellation underlying modus DARII in 9.6.2 may be depicted as follows:

9.6.3 SET-THEORETIC VIEW OF MODUS DARII

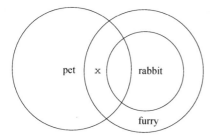

Premise 1 is expressed by the set rabbit being a subset of furry. Premise 2 is expressed by the set pet overlapping with the set rabbit. The conclusion is expressed by the set pet overlapping with the set furry.

DBS describes the set-theoretic constellation as follows:

9.6.4 REPHRASING MODUS DARII IN DBS

inference: α be_rabbit implies α be_furry.
   input: Some pets be_rabbit
  output: Some pets be_furry

The inference applies by binding *some pets* in the input to the variable α in the antecedent and using this binding in the consequent to derive the output. The input matches the antecedent and the consequent derives matching output.

Following standard procedure, this is shown in detail by the following translation of 9.6.4 into the data structure of DBS:

9.6.5 APPLYING MODUS DARII AS FORMALIZED IN DBS

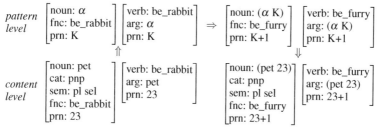

The particular affirmative quality of the judgment type **I**, i.e., the some, is coded by the features [cat: pnp] and [sem: pl sel] of the *pet* proplets at the content level. Because the grammatical properties of determiners are not reflected at the pattern level (compatibility by omission), modus DARII joins modus BARBARA and modus ponendo ponens as an instance of the DBS inference kind unnegated antecedent and unnegated consequent (9.2.6) in the variant *particular*.

We turn next to modus FERIO. The vowels in the name indicate the categorical judgment **E** (universal negative) in premise 1, **I** (particular affirmative) in premise 2, and **O** (particular negative) in the conclusion. In Predicate Calculus, this is formalized as follows:

9.6.6 MODUS FERIO IN PREDICATE CALCULUS

premise 1: $\neg\exists x[f(x) \wedge g(x)]$
premise 2: $\exists y[h(y) \wedge g(y)]$
conclusion: $\exists z[h(z) \wedge \neg g(z)]$

The middle term is g. If f is instantiated as is homework, g as is fun, and h as is reading, then the syllogism reads as follows:

9.6.7 INSTANTIATING MODUS FERIO

premise 1: There exists no x, x is homework and x is fun.
premise 2: For some y, y is reading and y is homework
conclusion: For some z, z is reading and z is no fun

The set-theoretic constellation underlying modus FERIO in 9.6.7 may be depicted as follows:

9.6.8 SET-THEORETIC VIEW OF MODUS FERIO

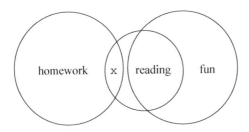

Premise 1 is shown by the sets homework and fun being disjunct. Premise 2 is depicted by the sets reading and homework overlapping. The conclusion is shown by the sets reading and homework, and reading and fun overlapping.

Consider the DBS inference schema for the set-theoretic constellation:

9.6.9 REPHRASING MODUS FERIO IN DBS

inference: α be_homework implies α not be_fun.
input: Some reading be_homework
output: Some reading not be_fun

The translation of 9.6.9 into the data structure of DBS is as follows:

9.6.10 APPLYING MODUS FERIO AS FORMALIZED IN DBS

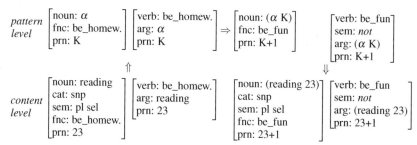

The content noun *some reading* is characterized by the features [cat: snp] and [sem: pl sel]. The particular negative quality of the judgment type **O** is coded by the feature [sem: *not*] in the *be_fun* proplets at the pattern as well as the content level. The reconstruction of FERIO in DBS joins the inferences of the kind unnegated antecedent and negated consequent (9.2.7) in the variant *particular*.

9.7 Modi BAROCO and BOCARDO

Like modus FERIO, modus BAROCO has the particular negative **O** in the conclusion. The **A** representing premise 1 indicates the categorical judgment universal affirmative (9.2.2).

9.7.1 MODUS BAROCO IN PREDICATE CALCULUS

premise 1: $\forall x[f(x) \rightarrow g(x)]$
premise 2: $\exists y[h(y) \wedge \neg g(y)]$
conclusion: $\exists z[h(z) \wedge \neg f(z)]$

The middle term is g. If f is instantiated as informative, g as useful, and h as website, then the syllogism reads as follows:

9.7.2 INSTANTIATING MODUS BAROCO

premise 1: All informative things are useful
premise 2: Some website are not informative
conclusion: Some websites are not useful

Among the classical syllogisms, BAROCO is special because the proof of its validity requires a reductio per impossibile.

The set-theoretic constellation underlying modus BAROCO in 9.7.2 may be depicted as follows:

9.7.3 SET-THEORETIC VIEW OF MODUS BAROCO

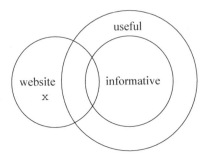

Premise 1 is shown by the set informative being a subset of useful. Premise 2 is depicted by the set website merely overlapping with the set informative. The conclusion is shown by the set website merely overlapping with useful.

The inference schema of DBS describes the set-theoretic constellation as follows:

9.7.4 REPHRASING BAROCO IN DBS

> inference: α not be_informative implies α not be_useful
> input: Some websites not be_informative
> output: Some websites not be_useful

Consider the translation of 9.7.4 into the data structure of DBS:

9.7.5 BAROCO IN DBS

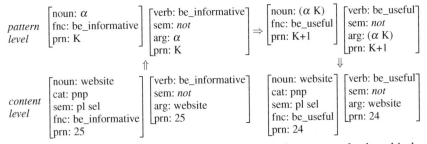

The reconstruction of BAROCO in DBS joins inferences of the kind negated antecedent and negated consequent (9.2.9) in the variant *particular*.

Like modus BAROCO, modus BOCARDO has the particular negative **O** in the conclusion. They differ in that the letters **A** and **O** in premises 1 and 2 are interchanged.

9.7.6 MODUS BOCARDO IN PREDICATE CALCULUS

> premise 1: $\exists x[f(x) \land \neg g(x)]$
> premise 2: $\forall y[f(y) \to h(y)]$
> conclusion: $\exists z[h(z) \land \neg g(z)]$

The middle term is f. If f is instantiated as be_cat, g as has_tail, and h as be_mammal, then the syllogism reads as follows:

9.7.7 INSTANTIATING MODUS BOCARDO

> premise 1: some cats have no tail
> premise 2: all cats are mammals
> conclusion: some mammals have no tail

The reductio per impossibile, which helps to prove the validity of BAROCO, is complemented in BOCARDO by ekthesis (Aristotle, An. Pr. I.6, 28b20–21).[10]

The set-theoretic constellation underlying modus BOCARDO in 9.7.7 may be depicted as follows:

9.7.8 SET-THEORETIC VIEW OF MODUS BOCARDO

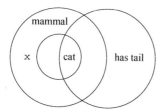

Premise 1 is the cat complement of the cat and has_tail intersection. Premise 2 is shown by cat being a subset of mammal. The conclusion is shown as the mammal complement of the mammal and has_tail intersection.

9.7.9 REPHRASING BOCARDO IN DBS

> inference: α be_cat implies some α not have_tail.
> input: some mammals are cats
> output: some mammals have no tail

Consider the translation of 9.7.8 into the data structure of DBS:

---

[10] In the middle ages, several jails in England, one specifically in Oxford, were called Bocardo because it was so hard for students to learn how to verify this syllogism.

9.7.10 APPLYING BOCARDO AS A DBS INFERENCE

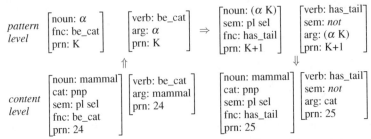

The subject in the consequent pattern of the BOCARDO inference is some α, a restriction which is coded by the feature [sem: pl sel], in contradistinction to the subject of the consequent pattern of the FERIO inference (9.6.9), which is unrestricted and thus compatible with universal, particular, and individual input (compatibility by omission)

9.8 Combining S- and C-Inferencing

Functional equivalence (CC 1.1, 15.1) at a certain level of abstraction between the human prototype and the artificial agent requires computational cognition to apply S- and C-inferencing in one and the same train of thought. Consider the following derivation of a data-driven countermeasure, which begins with the C-inferences 9.1.2 and CC 5.1.4, continues with a lexical S-inference coding a hypernymy (op. cit. 5.2.2), and concludes with another C-inference:

9.8.1 MIXING S- AND C-INFERENCE IN A TRAIN OF THOUGHT

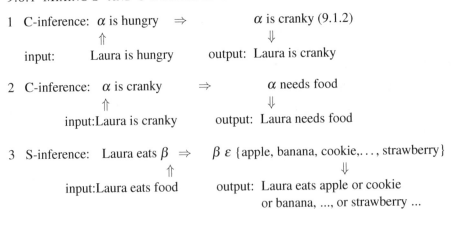

4 C-inference: α eats cookie $\Rightarrow$ α is agreeable again

　　　　　　　　　⇑　　　　　　　　　　⇓

　　　　　input:Laura eats cookie   output: Laura is agreeable again

The S-inference 3 illustrates a lexical alternative to the syllogisms analyzed in 9.3–9.7, namely a hypernymy, which is defined as follows.

9.8.2 LEXICAL S-INFERENCE IMPLEMENTING HYPERNYMY

[noun: α] $\Rightarrow$ [noun: β]
If α is animal, then β ε {ape, bear, cat, dog, ...}
If α is food, then β ε {apple, banana, cookie, ..., strawberry}
If α is fuel, then β ε {diesel, gasoline, electricity, hydrogen, ...}
　　　...

The set-theoretic structure of a hypernymy[11] is the relation between a superordinate term and its extension. Accordingly, food is the hypernym of apple, banana, cookie, ..., and strawberry. Set-theoretically, the denotation of food equals the codomain of α. The restrictions on variables are species-, culture-, and even agent-dependent. and may be approximated empirically by means of DBS corpus analysis (RMD[12] corpus).

9.9 Analogy

Common sense reasoning is based on relations provided by repeated observation and contingent knowledge. For example, there is nothing law-like or set-theoretic in Laura being cranky when hungry. There is another dimension, however, namely analogy: a truck not starting caused by a lack of fuel may be seen as analogous to being cranky caused by a lack of food.

9.9.1 COMMON SENSE REASONING BASED ON ANALOGY

1 C-inference: α has no fuel $\Rightarrow$ α does not start

　　　　　　　　　⇑　　　　　　　　　　⇓

　　　　　input: truck has no fuel   output: truck does not start

2 C-inference: α does not start $\Rightarrow$ α needs fuel

　　　　　　　　　⇑　　　　　　　　　　⇓

　　　　　input: truck does not start   output: truck needs fuel

3 S-inference: truck gets β $\Rightarrow$ β ε {diesel, gasoline, electricity,...}

　　　　　　　　　⇑　　　　　　　　　　⇓

　　　　　input: truck gets fuel   output: truck gets diesel or gasoline
　　　　　　　　　　　　　　　　　　　　　　or electricity or hydrogen...

---

[11] For the corresponding hyponymy see CC 9.1.1.
[12] Reference-Monitor corpus structured into Domains (CLaTR 15.3).

4   C-inference: α gets fuel $\Rightarrow$ α starts

$\Uparrow$ $\Downarrow$

input: truck gets fuel   output: truck starts

As in 9.9.1, the C-inference 1 is a general common sense observation, while the C-inference 2 applies to a particular instance. The lexical S-inference 3 is an instance of the hypernymy 9.8.2, while the C-inference 4 derives the desired result.

Using the data structure of DBS, the data-driven application of the second inference in 9.9.1 may be shown as follows:

9.9.2 APPLYING THE C-INFERENCE 2 OF 9.9.1

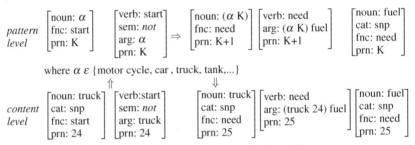

where $\alpha \; \varepsilon$ {motor cycle, car , truck, tank,...}

Finding and applying analogical countermeasures may be based in part on a systematic development of semantic fields (CC 11.3.3) across domains.

9.10 Conclusion

Based on the *categorical judgements* A, E, I, O, there are 256 (2^8) categorical syllogisms of which 24 have been found valid by the medieval scholastics. Using Euler's set-theoretical analysis of the categorical judgments, this chapter proposes to reanalyze the valid syllogisms as 12 DBS inferences. Depending on whether or not the antecedent, the consequent, or both are negated, there are 4 classes (9.2.6, 9.2.7, 9.2.8, 9.2.9) which are divided further by the distinction between a universal, particular, and individual variant.

Classical Barbara, Baroco, Bocardo, Celarent, Darii, Ferio, as well as modi ponendo ponens and tollendo tollens are reanalyzed in detail as DBS inferences. Two of the four classes are completed as (1) Barbara (universal, 9.5.4), Darii (particular, 9.6.4), and modus ponendo ponens (individual, 9.3.3), and (2) Celarent (universal, 9.5.9), Ferio (particular, 9.6.9), and modus tollendo tollens (individual, 9.4.2).

In human thought chains, S-inferences based on set theory and C-inferences based on common sense reasoning may be freely mixed. To support chaining, both kinds of inference use the same inference schema in agent-based DBS but differ in the source of reasoning, i.e., (i) set theoretic structures vs. (ii) something previously observed or learned by the agent.

10. Grounding of Concepts in Science

Mammalian cognition interacts with the raw data of its external and internal surroundings by means of concepts for shape, color, smell, taste, temperature, etc. (exteroception), but also hunger, thirst, and emotion (interoception). Recent advances in molecular biology and biochemistry define these concepts in terms of receptors consisting of complex proteins matching the input of characteristic molecules, but also in terms of protons in vision and sound waves in audition (*natural science*).

In parallel, driven by industrial applications, the biological mechanisms have been modeled artificially for some of these concepts, notably the electronic nose and the electronic tongue (*engineering science*). After showing why the approaches to meaning in philosophy and linguistics (*humanities*) are unsuitable for the grounding of concepts in the computational cognition of an autonomous robot, this chapter proposes to utilize the theoretical and practical advances of the neighboring sciences.

10.1 The Place of Concepts in a Content

DBS follows classical tradition in the humanities by distinguishing (i) three basic kinds of *contents*, (ii) three basic kinds of *concepts*, and (iii) two basic forms of *combination* (the big C's). The elementary contents are (1) name, (2) indexical, and (3) concept. The concepts are (a) referent, (b) relation, and (c) property (Sect. 4.1).

The combinations are the semantic relations of structure, i.e., (α) functor-argument and (β) coordination. In α, relations combine two or three referents into propositions, and properties modify referents, relations, or properties. In β, several referents, relations, or properties of the same kind concatenate into conjunctions.

Consider the interaction of content, concept, and relation in the complex language content of Lucy found a big blue square.:

10.1.1 FUNCTOR-ARGUMENT AND COORDINATION IN LANG. CONTENT

referent (noun)	relation (transitive verb)	property (adn)	property (adn)	referent (noun)
⌈sur: Lucy ⌉ noun: [**person x**] cat: snp sem: nm f fnc: **find** mdr: nc: pc: ⌊prn: 23 ⌋	⌈sur: found ⌉ verb: **find** cat: #n′ #a′ decl sem: ind past arg: [**person x**] **square** mdr: nc: pc: ⌊prn: 23 ⌋	⌈sur: big ⌉ adj: **big** cat: adn sem: pad mdd: **square** mdr: nc: **blue** pc: ⌊prn: 23 ⌋	⌈sur: blue⌉ adj: **blue** cat: adn sem: pad mdd: mdr: nc: pc: **big** ⌊prn: 23 ⌋	⌈sur: square ⌉ noun: **square** cat: snp sem: indef sg fnc: **find** mdr: **big** nc: pc: ⌊prn: 23 ⌋

© The Author(s), under exclusive license to Springer Nature Switzerland AG 2023
R. Hausser, *Ontology of Communication*, https://doi.org/10.1007/978-3-031-22739-4_10

A content is a set of proplets, defined as nonrecursive feature structures with ordered attributes, connected by the semantic relations of structure, coded by address.

The hear mode takes the surface Lucy found a big blue square. as input and derives the content 10.1.1 as output. The following derivation is (i) *surface-compositional* because each lexical item has a concrete sur value and there are no surfaces without a proplet analysis. The derivation order is (ii) *time-linear*, as shown by the stair-like addition of a next word proplet. The activation and application of operations is (iii) *data-driven* by input from automatic word form recognition:

10.1.2 TIME-LINEAR SURFACE-COMPOSITIONAL HEAR MODE DERIVATION

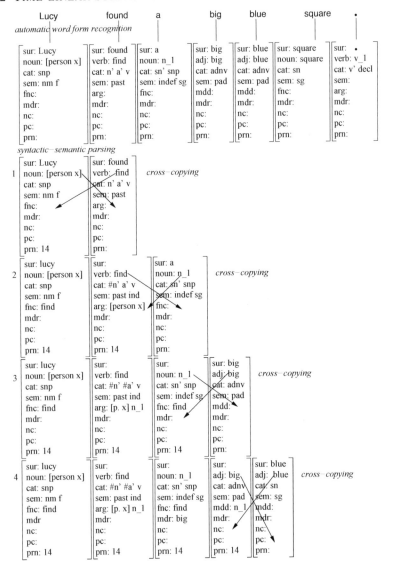

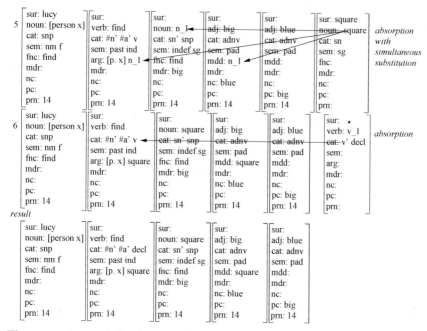

The operations of the hear mode use the connectives (1) × for cross-copying, (2) ∪ for absorption, and (3) ~ for suspension. Cross-copying encodes the semantic relations of structure such as SBJ×PRED (subject×predicate, line 2). Absorption combines a function word with a content word such as DET∪CN (line 1) or with another function word as in PREP∪DET (preposition∪determiner, CLaTR 7.2.5). Suspension such as ADV~NOM (TExer 3.1.3) applies if no semantic relation exists for connecting the next word with the content processed so far, as in Perhaps ~ Fido (slept.).

Each derivation step 'consumes' exactly one next word (reading). In a concatenation, the language-dependent sur value provided by lexical lookup is omitted.[1] Lexical lookup and syntactic-semantic concatenation are incrementally intertwined: lookup of a new next word occurs only after the current next word has been processed into the current sentence start.

In a graphical hear mode derivation like 10.1.2, cross-copying between two proplets is indicated by two diagonal arrows and the result is shown in the next line. This includes changes in the cat and the sem slots. For example, the canceling of the n' (nominative) and a' (accusative) valency positions in the cat slot of the *find* proplet of lines 2 and 3 is indicated by #-marking.[2]

---

[1] A partial exception are name proplets, which preserve their sur value in the form of a marker written in lower case default font, e.g., lucy. In the speak mode, the marker is converted back into a regular sur value written in Helvetica, e.g., Lucy.

[2] Canceling by #-marking preserves the canceled value for use in the DBS speak mode. This is in contradistinction to Categorial Grammar (CG), which cancels valency positions by deletion (loss of information). As a sign-based system, CG does not distinguish between the speak and the hear mode.

The speak mode takes a content like 10.1.1 as input and produces a language-dependent surface as output. Graphically, the semantic relations of functor-argument are represented as the connectives / for subject/predicate, \ for object\predicate, and | for modifier|noun, modifier|verb, and modifier|modifier. The semantic relations of coordination are represented graphically as the connective (a) − for noun−noun, (b) verb−verb, (c) adn−adn, and (d) adv−adv.

Based on the definition of graphical /, \, |, and − for the semantic relations of structure, DBS analyzes a content like 10.1.1 in four standard views:

10.1.3 SEMANTIC RELATIONS UNDERLYING SPEAK MODE DERIVATION

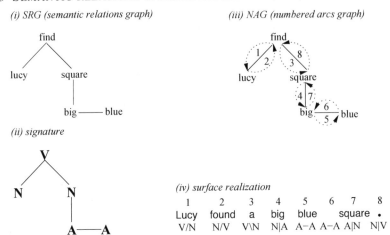

The (i) SRG (*semantic relations graph*) uses the core values lucy, find, square, big and blue of 10.1.1 as nodes. The (ii) *signature* uses the core attributes N(oun), V(erb), and A(dj) as nodes. The (iii) NAG (*numbered arcs graph*) completes the SRG with traversal numbers and shows content activation by the time-linear navigation through the semantic hierarchy in the think mode. The traversal numbers are used in the (iv) *surface realization*; it optionally realizes language-dependent surfaces in a speak mode which rides piggyback on the think mode navigation.

In summary, the speak mode's time-linear navigation through the input content (10.1.1) achieves a *linearization* of a semantic hierarchy into a language surface which is suitable for an incremental content transfer from speaker to hearer. The hear mode's surface-compositional derivation (10.1.2) achieves a re-*hierarchization* of the speaker's content from the time-linear input surface.

10.2 Definition of Concepts at the Elementary, Phrasal, or Clausal Level?

In a syntactic-semantic analysis of a content like 10.1.3, the concepts (as the elementary building blocks of DBS cognition) are shown by placeholder values, using English base forms for convenience. Let us turn now to the topic of this chapter, namely the question of how to complete place holder values with computational implementations suitable for the cognition of an artificial autonomous agent.

A rough semantic delimitation of a place holder concept is provided by its attribute in a proplet. The concept values of the core attribute noun and the continuation attribute arg are restricted to an *argument*. The concept values of the core attribute verb and the continuation attribute fnc are restricted to a one-, two-, or three-place *functor*. The concept values of the core attribute adj and the continuation attribute mdr are restricted to an adnominal or adverbial *modifier*. When using the semantic notion of 'property', the modifiers and the one-place verbs are *properties*, and the two- and three-place verbs are *relations* (CC 1.5).

For the computational cognition of an autonomous robot, the semantics must be complemented by an interface component with sensors for recognition and actuators for action. This requires the cooperation of all three branches of today's science, i.e., (a) the natural sciences, (b) the engineering sciences, and (c) the humanities.

Consider, for example, the type of the concept *blue* and an associated token:

10.2.1 ELEMENTARY MEANING ANALYSIS: CONCEPT DEFINITION

type
$$\begin{bmatrix} \text{place holder: blue} \\ \text{sensory modality: vision} \\ \text{semantic field: color} \\ \text{content kind: concept} \\ \text{wavelength: 450–495nm} \\ \text{frequency: 670–610 THz} \\ \text{samples: a, b, c, ...} \end{bmatrix}$$

token
$$\begin{bmatrix} \text{place holder: blue} \\ \text{sensory modality: vision} \\ \text{semantic field: color} \\ \text{content kind: concept} \\ \text{wavelength: 470nm} \\ \text{frequency: 637 THz} \end{bmatrix}$$

In the type, the values of wavelength and frequency are intervals, but constants in the token.

The humanities provide the type-token distinction in philosophy (Peirce 1906, CP Vol.4, p. 375), which goes back to the distinction between the necessary and the accidental (Aristotle, Metaphysics, Books ζ and η); the natural sciences provide the wavelength interval 450–495nm in the type; and the engineering sciences provide the wavelength measurement 470nm in the token.

Alternative to an analysis at the elementary level of grammatical complexity, there are long standing proposals in the humanities which analyze meaning at the phrasal and at the clausal level. Unlike 10.2.1, they fail to fit into the core and continuation slots of a DBS content (10.1.1), but for the purpose of building a talking robot they have other disqualifying properties worth noting.

The most widely used meaning analysis is at the phrasal level and based on informally paraphrasing a definiendum with a definiens:

10.2.2 PHRASAL MEANING ANALYSIS: PARAPHRASE

blue = the color of the cloudless sky

The definiendum is nominalized blue and the definiens the phrasal noun the color of the cloudless sky.[3] To be meaningful, the hearer must understand the words of the definiens and the semantics of their composition. Within a language community, this kind of meaning explanation is most effective. For a talking robot under hard- and software construction, however, it is unsuitable because the definition of a word in terms of other words runs into the problem of circular paraphrasing.[4]

Circular paraphrasing is avoided by a meaning analysis at the clausal level, namely the definition of a formal metalanguage (Tarski 1935, 1944):

10.2.3 CLAUSAL MEANING ANALYSIS: META-LANGUAGE DEFINITION

'der Himmel ist blau' is a true sentence if and only if the sky is blue.

To avoid logical inconsistency, the metalanguage must be (i) formally constructed, (ii) its notions must be mathematically obvious (such as set relations), and (iii) the object language may not contain the truth predicates true and false (FoCL 19–21). For building a talking robot, the first and the second condition are impractical, and the third makes the solution incomplete.

10.3 Extending a Concept to Its Class

The explicit concept definition 10.2.1 may be generalized routinely to other colors:

10.3.1 SIMILARITY AND DIFFERENCE BETWEEN COLOR CONCEPT TYPES

$$
\begin{bmatrix}
\text{place holder: red} \\
\text{sensory modality: vision} \\
\text{semantic field: color} \\
\text{content kind: concept} \\
\text{wavelength: 700-635 nm} \\
\text{frequency: 430-480 THz} \\
\text{samples: a, b, c, ...}
\end{bmatrix}
\begin{bmatrix}
\text{place holder: green} \\
\text{sensory modality: vision} \\
\text{semantic field: color} \\
\text{content kind: concept} \\
\text{wavelength:495-570 nm} \\
\text{frequency: 526-606 THz} \\
\text{samples: } a', b', c', ...
\end{bmatrix}
\begin{bmatrix}
\text{place holder: blue} \\
\text{sensory modality: vision} \\
\text{semantic field: color} \\
\text{content kind: concept} \\
\text{wavelength: 490-450 nm} \\
\text{frequency: 610-670 THz} \\
\text{samples: } a'', b'', c'', ...
\end{bmatrix}
$$

The three types differ in their wavelength and frequency intervals, and their place holder and samples values; they share the sensory modality, semantic field, and content kind values.

Computationally, the use of a DBS concept in recognition is based on matching the type pattern with raw data, resulting in an instantiating token:

10.3.2 RECOGNITION OF blue

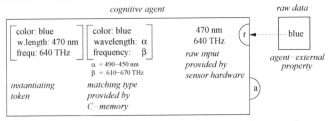

In action, the use of a DBS concept is based on adapting the type to a purpose, resulting in a token, realized by the agent's interface component as raw data:

---

[3] The example is from https://www.thefreedictionary.com/blue.

[4] Noted by de Saussure ([1916]1972) and explicated further by D. Lewis (1969).

10.3.3 ACTION OF REALIZING blue

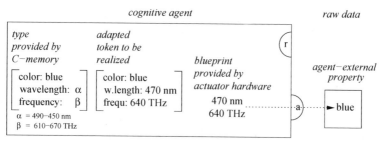

An example is a cuttlefish (metasepia pfefferi) using its chromatophores. Another place holder value for a concept in 10.1.1 is square:

10.3.4 TYPE AND TOKEN OF THE CONCEPT square

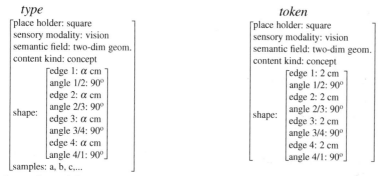

The edge value of the type is a variable which matches an infinite number of square tokens with different edge lengths.

Just as the definition of the concept *blue* may be generalized routinely to other colors (10.3.1), the definition of the concept *square* may be generalized to other shapes in two-dimensional geometry, such as *equilateral triangle*, and *rectangle*:

10.3.5 SIMILARITY AND DIFFERENCE BETWEEN CONCEPT SHAPE TYPES

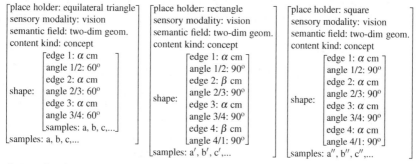

Most abstractly, the recognition of a square may be shown as follows:

10.3.6 USING A CONCEPT TYPE IN NONLANGUAGE RECOGNITION

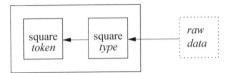

The raw input data are provided by the agent's interface component. They are recognized as an instance of the two-dimensional shape square because there are four lines of equal length and the angle of their intersections is 90°.

The type of square may also be used in action, as when drawing a square:

10.3.7 USING A CONCEPT TOKEN IN NONLANGUAGE ACTION

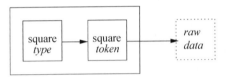

The definition of concept types, corresponding concept tokens, and raw data relies on the natural sciences, here geometry. Type-token matching in recognition and action is an instance of computational pattern matching in DBS (CC 1.6).

10.4 Language Communication

The speak mode is the language variant of nonlanguage action. It re-uses type-token adaptation for the production of language-dependent surfaces. In the medium of speech, a surface token differs from its type by specifying volume, pitch, speed, timbre, etc., and in the medium of writing by specifying font, size, color, etc., i.e., what Aristotle would call the accidental properties.

The hear mode is the language variant of nonlanguage recognition. It re-uses token-type matching for assigning language-dependent types to raw surface data. Re-use of earlier mechanisms in the evolutionary transition from nonlanguage to language cognition is in the spirit of Charles Darwin and out of reach for theories based on a sign-based substitution-driven ontology.

Because the transfer of content is based on raw data, (a) the concept types, (b) the language dependent surface types, and (c) the conventions connecting (a) and (b) exist solely in the respective cognitions of speaker and hearer (anything else would be reification). Successful communication presupposes that speaker and hearer have learned the same natural language. In addition, the speaker must be able to produce surface types as tokens and the hearer must be able to recognize surface tokens by means of matching types.

The extension of type-token matching from nonlanguage cognition to language cognition may shown schematically as follows:

10.4.1 COMBINING NONLANGUAGE INTO LANGUAGE COGNITION

(i) nonlanguage action   nonlanguage recognition

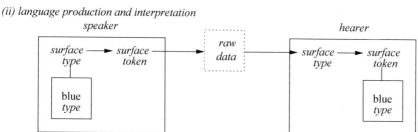

(ii) language production and interpretation

Graph (i) shows activation and recognition of the color blue, possibly by the same agent, e.g., metasepia pfefferi, at different occasions.

Graph (ii) explains the transfer of content with raw data, e.g., sound waves or pixels, from a speaker to a hearer. The language-dependent surfaces have no meaning or grammatical properties and their meaning₁ exists solely in the agents' cognition.

In the medium of writing, the type-token adaptation of a surface production may be illustrated in more detail as follows:

10.4.2 SPEAK MODE: FROM CONTENT TO SURFACE TYPE TO RAW DATA

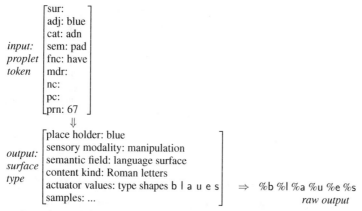

The input, i.e., the proplet token *blue* of nonlanguage cognition, retrieves the corresponding language-dependent surface, here the type of German b l a u e s from the agent's memory, based on a list which provides allomorphs using the input proplet's core, cat, and sem values, and the rules of morphological composition (FoCL 13, 14).

The type shapes of this output serve as input to an actuator of the agent's interface component which adapts the surface type into a token and realizes it as raw data.

The type-token instantiation in the corresponding surface recognition may be illustrated in more detail as follows:

10.4.3 HEAR MODE: FROM RAW DATA TO SURFACE TYPE TO CONTENT

pattern: surface type
$$
\begin{bmatrix}
\text{place holder: blue} \\
\text{sensory modality: vision} \\
\text{semantic field: language surface} \\
\text{content kind: roman letters} \\
\text{shape types: b l a u e s} \\
\text{samples: ...}
\end{bmatrix}
\Rightarrow
$$

output: surface token
$$
\begin{bmatrix}
\text{place holder: blue} \\
\text{sensory modality.: vision} \\
\text{semantic field: language surface} \\
\text{content kind: roman letters} \\
\text{sensor values: shape tokens b\% l\% a\% u\% e\% s\%} \\
\text{samples: ...}
\end{bmatrix}
$$

⇑
sensory modality: vision
input: raw data

The input consists of raw data which are provided by vision sensors in the agent's interface component and matched by the letters' shape types provided by the agent's memory. The output replaces the shape types, here b l a u e s, with the matching raw data resulting in shape tokens. Shown as b% l% a% u% e% s%, they record the accidental properties. The value crucial for the understanding of the hearer is the place holder, here *blue*, for the lexical look-up of the correct nonlanguage concept.

10.5 Combining Concepts Into Content

At this point, we have explicit definitions of the three color concepts red, green, and blue (10.3.1), and the three shape concepts triangle, rectangle, and square (10.3.5). Instantiated as tokens, they combine into nine two-concept contents.

10.5.1 Three out of nine two-concept contents

$$
\begin{bmatrix}
\text{sur:} \\
\text{noun: red} \\
\text{cat: adnv} \\
\text{sem: pad} \\
\text{mdd: triangle} \\
\text{mdr:} \\
\text{nc:} \\
\text{pc:} \\
\text{prn: 23}
\end{bmatrix}
\begin{bmatrix}
\text{sur:} \\
\text{noun: triangle} \\
\text{cat: sn} \\
\text{sem: sg} \\
\text{fnc:} \\
\text{mdr: red} \\
\text{nc:} \\
\text{pc:} \\
\text{prn: 23}
\end{bmatrix}
\begin{bmatrix}
\text{sur:} \\
\text{noun: green} \\
\text{cat: adnv} \\
\text{sem: pad} \\
\text{mdd: rectangle} \\
\text{mdr:} \\
\text{nc:} \\
\text{pc:} \\
\text{prn: 24}
\end{bmatrix}
\begin{bmatrix}
\text{sur:} \\
\text{noun: rectangle} \\
\text{cat: sn} \\
\text{sem: sg} \\
\text{fnc:} \\
\text{mdr: green} \\
\text{nc:} \\
\text{pc:} \\
\text{prn: 24}
\end{bmatrix}
\begin{bmatrix}
\text{sur:} \\
\text{noun: blue} \\
\text{cat: adnv} \\
\text{sem: pad} \\
\text{mdd: square} \\
\text{mdr:} \\
\text{nc:} \\
\text{pc:} \\
\text{prn: 25}
\end{bmatrix}
\begin{bmatrix}
\text{sur:} \\
\text{noun: square} \\
\text{cat: sn} \\
\text{sem: sg} \\
\text{fnc:} \\
\text{mdr: blue} \\
\text{nc:} \\
\text{pc:} \\
\text{prn: 25}
\end{bmatrix}
$$

Tweaking the core and continuation values in the tokens results in infinitely (theoretically, using real numbers $\mathbf{R}$) many different contents. If the values are kept constant, adding a new color and a new shape increases the number of resulting contents polynomially: 1·1, 2·2, 3·3, 4·4, etc. If language-dependent surface types are added, the number of instantiations depends on the number of languages.

The definitions of the color types (10.3.1) and tokens are suitable for robotic cognition because they are represented by numbers which have interpretations as measurements in the color spectrum. It is similar for the definition of shape types (10.3.5) and tokens because they are represented by numbers which have interpretations as measurements of line lengths and angle degrees in two-dimensional geometry.

This is different from, for example, Bjørner's (1978) program for keeping track of a grocery store's inventory:

> p. 34
> 'A grocery is here selectively abstracted by abstractions of its shelves and store, i.e., inventory, its cash register, and its catalogue.'

The program allows inferencing such as the following:

> pp. 36/37
> 'ii. If the ware additionally is further stored in the back room, then the number of items on the shelves must actually fall between the minimum, lower and maximum, upper bounds;'

Relying on human understanding of the English words, the program is suitable for an English speaking human grocer to keep track of the inventory. An autonomous robot, however, requires an interpretation of the concepts based on the natural and engineering sciences. Bjørner's program is not a metalanguage in the sense of Tarski (10.2.3), but a procedural implementation of some numerical aspects abstracted from the subject matter.

10.6 Language Surfaces and Meaning₁ Concepts in Communication

Natural language communication requires the use of two different but *complementary modalities*, one for the speak and the other for the hear mode, whereby speaker and hearer are different individuals, with the exception of soliloquy. For example, in the medium of speech the sensory modality of vocalization must be used for action (surface production in the speak mode) and the complementary modality of audition for recognition (surface interpretation in the hear mode). In the medium of writing the sensory modality of manipulation must be used for action and the complementary modality of vision for recognition, and accordingly for Braille and signing.

The two complementary modalities (tcm) requirement is limited to cooperative behavior, specifically surface processing in natural language communication (12.8). In other areas of cognition, an individual may freely select one or more single modalities (sm). For example, in manipulation, an individual may button a shirt (action) without looking (recognition). In vision, an individual may be bird watching (recognition) without moving (action). In short, for using several modalities simultaneously, they must be compatible but there is no requirement of being complementary, as when eating while watching tv or talking while driving.

In communication, (i) the complementary modalities of surface concepts needed for content transfer between individuals and (ii) the contents of nonlanguage cognition serving as literal meanings₁ are inextricably connected by convention in the language community:

10.6.1 MEANING₁ TRANSFER BY MEANS OF A SURFACE TOKEN

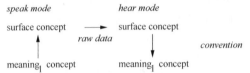

In addition to language communication and nonlanguage recognition and action, meaning₁ may be used abstractly in reasoning as content *per se*.

10.7 Extero- and Interoception

Research in molecular biology and biochemistry explains the cognitive mechanism of concepts in terms of complex receptor proteins which match specific kinds of molecules, resulting in the associated sensation. Driven by practical needs in industry, this biological mechanism has been recreated artificially in some instances.

For example, the taste concepts sweet, sour, salty, bitter, and umami (savoriness) have been modeled as an 'electronic tongue' (Winquist 2008). In pharmaceutics and the food-beverage sector, these artificial receptors are used to analyze flavor ageing in beverages, quantify taste masking efficiency of formulations, and monitor biological and biochemical processes.

Also of interest for industry are odors and flavors, which led to the rapid development of an 'electronic nose.' Persaud and George (1982) tuned their artificial receptor to an axis ranging from very pleasant (rose) to very unpleasant (skunk). It appeared that odorant pleasantness is tightly linked to molecular structure (Haddad et al. 2010). Therefore, the concepts of smell, like those of taste, may be based on molecules (raw input data) which are matched by receptor proteins (type).

The technology of the electronic tongue and nose satisfies the theoretical and practical necessity of grounding concepts in DBS. It suggests the construction of artificial receptors also for other domains.

Natural receptors at the periphery of the agent's body, as for taste and smell, are called exteroceptors. They include vision, hearing, touch, and thermoception, i.e., the feeling of hot and cold. Also called the conductive modality (Filingeri 2016), termoception is based on receptors in the skin.

Receptors inside the agent's body are called interoceptors (Connell et al. 2017). Triggered by body-internal deviations from a state of homeostasis, they are responsible for the 'drive states,' e.g., hunger and thirst. For example, a low glucose level in the blood stream is recognized by receptors in the brain (Rolls 2000), interpreted as hunger, and countered by raising the glucose level to the set point by ingesting food (negative feedback in control theory, Wiener 1948).

In addition to the concepts for physiological states, there are the concepts which recognize psychological states such as fatigue, vigor, relaxation, and boredom. Of these, fatigue has drawn special attention in medicine because of post-infectious fatigue (Kazuhiro Kondo 2006), but also because of the Chronic Fatigue Syndrome (CFS), including the Gulf War Syndrome.

10.8 Emotion

In the humanities, emotion has been intensely researched from an outside observer's point of view. Ekman (1999, p. 46) explains the function of emotion as a pathway for fast, comprehensive access 'to deal quickly with important interpersonal encounters,' like running for dear life triggered by the emotion of fear. Rimé (2009) describes the social mechanisms of sharing emotions. Lerner et al. (2015) investigate the influence of emotions on decision making, specifically in business.

But what about the emotion concepts, such as anger, surprise, fear, disgust, happiness, and sadness? They are also based on protein receptors, here in the brain, matching certain molecules, here in the blood stream, which explains why 'emotions are unbidden, not chosen by us' (Ekman op. cit. p. 54).

10.9 Conclusion

The ultimate standard of verification for the DBS theory of language is the construction of a robot capable of communicating freely in natural language. The method is incremental upscaling of a declarative specification with an operational implementation for the automatic evaluation of systematic test scenarios.

To achieve this standard, the century old division (Snow [1959] 2001) between the humanities, on the one hand, and the natural and the engineering sciences, on the other, must be overcome. It is as much a need for connecting the sciences to the humanities as for connecting the humanities to the sciences.

A case in point is the treatment of elementary meanings in the humanities, specifically philosophy and linguistics, on the one hand, and the natural and engineering sciences, specifically molecular biology and biochemistry, on the other. Written from the humanities' perspective of computational linguistics, it is suggested that the sciences provide a grounding of concepts suitable for building a talking robot.

11. Function Words

The vocabulary of a natural language is divided into content words (autosemantica) like book or read, and function words like the or and (synsemantica).[1] Examples of content word categories are noun, verb, and adj, those of function words determiner, preposition, auxiliary, and conjunction.

Typologically, isolating languages like English and Chinese prefer function words and word order for coding semantic relations within and between noun, verb, and adj contents, while inflectional languages like classical Latin and agglutinating languages like Korean prefer morphology, i.e., affixes attached to content word surfaces.

This chapter concentrates on the grammatical role of function words in English, and compares it with corresponding constructions in a language which uses more morphology than English, i.e., German. In line with the agent-based data-driven ontology of DBS, the syntactic-semantic mechanism of function words is shown in the hear and speak mode.

11.1 Introduction

Natural languages differ in the way in which complex contents are coded. For example, in classical Latin the partial content pro1 see' has the single surface video, but in English the two surfaces I see. The following DBS analyses show what the two codings have in common and where they differ:

11.1.1 DBS PROPLET PRESENTATION OF I see IN LATIN AND ENGLISH

Latin: morphology   English: syntactoc-semantic composition (cross-copying $\mathbf{SUBJ} \times \mathbf{PRD}$)

$$
\begin{bmatrix}
\text{sur: video} \\
\text{verb: see} \\
\text{cat: \#s1' a' v} \\
\text{sem: pres ind} \\
\text{arg: pro1} \\
\text{mdr:} \\
\text{nc:} \\
\text{pc:} \\
\text{prn: 93}
\end{bmatrix}
\begin{bmatrix}
\text{sur: I} \\
\text{noun: \textbf{pro1}} \\
\text{cat: snp} \\
\text{sem: s1} \\
\text{fnc:} \\
\text{mdr:} \\
\text{nc:} \\
\text{pc:} \\
\text{prn: 93}
\end{bmatrix}
\begin{bmatrix}
\text{sur: see} \\
\text{verb: \textbf{see}} \\
\text{cat: n-s3' a' v} \\
\text{sem: pres ind} \\
\text{arg:} \\
\text{mdr:} \\
\text{nc:} \\
\text{pc:} \\
\text{prn}
\end{bmatrix}
\Rightarrow
\begin{bmatrix}
\text{sur:} \\
\text{noun: \textbf{pro1}} \\
\text{cat: snp} \\
\text{sem: s1} \\
\text{fnc: \textbf{see}} \\
\text{mdr:} \\
\text{nc:} \\
\text{pc:} \\
\text{prn: 93}
\end{bmatrix}
\begin{bmatrix}
\text{sur:} \\
\text{verb: \textbf{see}} \\
\text{cat: \#n-s3' a' v} \\
\text{sem: pres ind} \\
\text{arg: \textbf{pro1}} \\
\text{mdr:} \\
\text{nc:} \\
\text{pc:} \\
\text{prn: 93}
\end{bmatrix}
$$

In Latin, the surface and its syntactic-semantic content are selected from the verbal paradigm of the inflectional morphology. It provides variations of person, number.

---

[1] Marty 1918, pp.205 ff.

R. Hausser, *Ontology of Communication*, https://doi.org/10.1007/978-3-031-22739-4_11

tense, and verbal mood, e.g., vides, videam, videbam, viderem. In English, in contrast, two content proplets with the surfaces I and see are connected by the cross-copying operation **SBJ**×**PRD** of the hear mode. For variations of verbal mood and tense other than indicative present, English uses function words, e.g., have/has seen or could have seen. The grammatical objects, in contrast, i.e., te in Latin and you in English, are treated alike in the two languages, namely by syntactic-semantic composition: Te video[2] and I see you.

In addition to *affixing* (morphological composition) in regular nouns (e.g., book, book+s), verbs (e.g., correct, correct+ed), and adjs (e.g., fast, fast+er, fast+est), there is *allomorphy*, i.e., variation of the word stem (FoCL 13). Examples of English allomorphy are the nouns foot, feet; mouse, mice, the verbs see, saw, seen; buy, bought, bought, and the adj good, better, best (suppletion).

For syntactic-semantic composition, the analyses of grammatically corresponding regular and irregular forms are coded alike (proplet normalization):

11.1.2 REGULAR VS. IRREGULAR VERB FORMS IN ENGLISH

regular verb form

```
⎡sur: correct+ed⎤
⎢verb: correct    ⎥
⎢cat: n′ a′ v     ⎥
⎢sem: past ind    ⎥
⎢arg:             ⎥
⎢mdr:             ⎥
⎢nc:              ⎥
⎢pc:              ⎥
⎣prn              ⎦
```

irregular verb form

```
⎡sur: saw      ⎤
⎢verb: see     ⎥
⎢cat: n′ a′ v  ⎥
⎢sem: past ind ⎥
⎢arg:          ⎥
⎢mdr:          ⎥
⎢nc:           ⎥
⎢pc:           ⎥
⎣prn           ⎦
```

The regular and the irregular verb form share corresponding positions in their respective paradigms and their proplets differ only in the sur and core values. The empty slots are used by syntactic-semantic composition.

Proplet normalization may also be applied between different but typologically similar languages, as shown by the following English_German counterparts correct+ed_ korrigier+te (both regular) and saw_sah (both irregular):

11.1.3 CORRESPONDING FORMS IN ENGLISH AND GERMAN

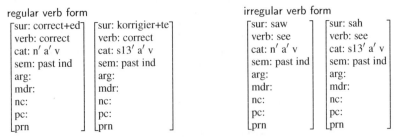

regular verb form

```
⎡sur: correct+ed⎤  ⎡sur: korrigier+te⎤
⎢verb: correct    ⎥  ⎢verb: correct      ⎥
⎢cat: n′ a′ v     ⎥  ⎢cat: s13′ a′ v     ⎥
⎢sem: past ind    ⎥  ⎢sem: past ind      ⎥
⎢arg:             ⎥  ⎢arg:               ⎥
⎢mdr:             ⎥  ⎢mdr:               ⎥
⎢nc:              ⎥  ⎢nc:                ⎥
⎢pc:              ⎥  ⎢pc:                ⎥
⎣prn              ⎦  ⎣prn                ⎦
```

irregular verb form

```
⎡sur: saw      ⎤  ⎡sur: sah       ⎤
⎢verb: see     ⎥  ⎢verb: see      ⎥
⎢cat: n′ a′ v  ⎥  ⎢cat: s13′ a′ v ⎥
⎢sem: past ind ⎥  ⎢sem: past ind  ⎥
⎢arg:          ⎥  ⎢arg:           ⎥
⎢mdr:          ⎥  ⎢mdr:           ⎥
⎢nc:           ⎥  ⎢nc:            ⎥
⎢pc:           ⎥  ⎢pc:            ⎥
⎣prn           ⎦  ⎣prn            ⎦
```

In other respects, the proplet definitions of English-German counterparts may diverge. For example, German noun proplets require grammatical gender specification for determiner+noun agreement, which would not be appropriate for English.

11.2 Interpreting Determiner Noun Combination in Hear Mode

A syntactic-semantic operation of the DBS hear mode combines a sentence start with a next word. There are three kinds of functor-argument[3] combination: (i) *cross-copying* between two proplets (connective $\times$), (ii) *absorption* of a content word into a function word (connective $\cup$), and (iii) *suspension* when an application has to be postponed because the word form to be connected with has not yet arrived (connective $\sim$).

The absorption of a content word into a function word may be shown by the following application of the hear mode operation **DET∪CN**:

11.2.1 PLURAL DETERMINER+NOUN COMPOSITION IN ENGLISH

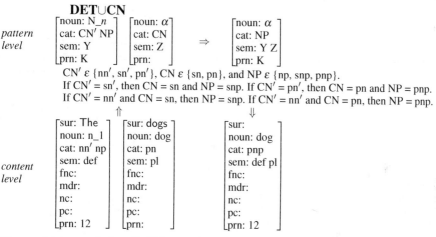

DET∪CN

pattern level

$$\begin{bmatrix} \text{noun: N\_}n \\ \text{cat: CN}'\ \text{NP} \\ \text{sem: Y} \\ \text{prn: K} \end{bmatrix} \begin{bmatrix} \text{noun: } \alpha \\ \text{cat: CN} \\ \text{sem: Z} \\ \text{prn:} \end{bmatrix} \Rightarrow \begin{bmatrix} \text{noun: } \alpha \\ \text{cat: NP} \\ \text{sem: Y Z} \\ \text{prn: K} \end{bmatrix}$$

CN$'$ ε {nn$'$, sn$'$, pn$'$}, CN ε {sn, pn}, and NP ε {np, snp, pnp}.
If CN$'$ = sn$'$, then CN = sn and NP = snp. If CN$'$ = pn$'$, then CN = pn and NP = pnp.
If CN$'$ = nn$'$ and CN = sn, then NP = snp. If CN$'$ = nn$'$ and CN = pn, then NP = pnp.

content level

$$\begin{bmatrix} \text{sur: The} \\ \text{noun: n\_1} \\ \text{cat: nn}'\ \text{np} \\ \text{sem: def} \\ \text{fnc:} \\ \text{mdr:} \\ \text{nc:} \\ \text{pc:} \\ \text{prn: 12} \end{bmatrix} \begin{bmatrix} \text{sur: dogs} \\ \text{noun: dog} \\ \text{cat: pn} \\ \text{sem: pl} \\ \text{fnc:} \\ \text{mdr:} \\ \text{nc:} \\ \text{pc:} \\ \text{prn:} \end{bmatrix} \begin{bmatrix} \text{sur:} \\ \text{noun: dog} \\ \text{cat: pnp} \\ \text{sem: def pl} \\ \text{fnc:} \\ \text{mdr:} \\ \text{nc:} \\ \text{pc:} \\ \text{prn: 12} \end{bmatrix}$$

The variable restriction If CN$'$ = sn$'$, then CN = sn and NP = snp ensures that a singular determiner must take a singular noun argument, e.g., a dog and every dog. The restriction If CN$'$ = pn$'$, then CN = pn and NP = pnp ensures that a plural determiner must take a plural noun argument, e.g., all dogs. In both, it is the determiner (functor) which determines the grammatical number of the result.

The restriction If CN$'$ = nn$'$ and CN = sn, then NP = snp ensures that a definite determiner and a singular noun result in a singular noun phrase, e.g., the dog. The restriction If CN$'$ = nn$'$ and CN = pn, then NP = pnp ensures that a definite de-

---

[2] The choice between morphology and syntax occurs also within a language: awaiting the decision vs. waiting for the decision. A language may use a function word and an affix, e.g., Latin et and -que, for the same meaning, i.e., *and*.

[3] For coordination see 11.7.

terminer and a plural noun result in a plural noun phrase, e.g., the dogs. Here it is the noun (argument) which determines the grammatical number of the result.[4]

That a dog and the dog denote a single individual and all dogs, the dogs as well as every dog denote plural sets is coded lexically as the sem value of the determiner proplet. The lexical properties of the English determiners and the variable restrictions of the hear mode operation 11.2.1 result in the following proplets:

11.2.2 PROPLETS OF a dog, the dog, every dog, all dogs, AND the dogs

⌈sur: a dog ⌉	⌈sur: the dog⌉	⌈sur: every dog⌉	⌈sur: all dogs⌉	⌈sur: the dogs⌉
noun: dog	noun: dog	noun: dog	noun: dog	noun: dog
cat: snp	cat: snp	cat: snp	cat: pnp	cat: pnp
sem: indef sg	sem: def sg	sem: pl	sem: indef pl	sem: def pl
fnc:	fnc:	fnc:	fnc:	fnc:
mdr:	mdr:	mdr:	mdr:	mdr:
nc:	nc:	nc:	nc:	nc:
pc:	pc:	pc:	pc:	pc:
⌊prn: 12 ⌋	⌊prn: 12 ⌋	⌊prn: 12 ⌋	⌊prn: 12 ⌋	⌊prn: 12 ⌋

The noun phrases a dog and the dog share the cat value snp and the sem value sg, but differ in the sem values indef and def. All dogs and the dogs share the cat value pnp and the sem value pl, but differ in the sem values indef and def. Every dog and all dogs share the sem value pl but differ in the cat values snp and pnp.

The German counterparts to the English examples in 11.2.2 are defined as follows:

11.2.3 PROPLETS OF ein Hund, der H., jeder H., alle Hunde, die Hunde

⌈sur: ein Hund[5]⌉	⌈sur: der Hund⌉	⌈sur: jeder Hund⌉	⌈sur: alle Hunde⌉	⌈sur: die Hunde⌉
noun: dog	noun: dog	noun: dog	noun: dog	noun: dog
cat: s3 m	cat: s3 m	cat: s3 m	cat: p3	cat: p3
sem: indef sg	sem: def sg	sem: pl	sem: indef pl	sem: def pl
fnc:	fnc:	fnc:	fnc:	fnc:
mdr:	mdr:	mdr:	mdr:	mdr:
nc:	nc:	nc:	nc:	nc:
pc:	pc:	pc:	pc:	pc:
⌊prn: 12 ⌋	⌊prn: 12 ⌋	⌊prn: 12 ⌋	⌊prn: 12 ⌋	⌊prn: 12 ⌋

The definite article the in English has only one form for singular and plural, while the definite article in German has the forms der, die, das, des, dem, den for coding case, number, and gender.

Case is needed for filling the correct valency slot of the predicate. Number is needed for the nominative, as in der Hund bellte vs. die Hunde bellten. Gender is needed in the singular for coreference with a possible personal pronoun, as in die Frau...sie

---

[4] The asymmetry between English indefinite and definite determiners regarding the source of grammatical number may be a problem for the *head-dependent* distinction (Osborne&Maxwell 2015) in Dependency Grammar (Mel'čuk 1988), but not for the semantically neutral notions of functor (slot) and argument (filler).

[5] As non-elementary contents, the sur slots in 11.2.2 and 11.2.3 would normally be empty, but are used here for convenience.

or ihr.[6] The differentiated determiner+noun combinations of German regarding case, number, and gender require variable restrictions which are substantially different from English and constitute a challenge for translating from English to German.

11.3 Producing Determiner Noun Combination in Speak Mode

As a minimal requirement for successful language communication, the content used as input to the speak mode and the content produced as output of the hear mode must be the same. To show a content *per se*, DBS uses two formats. One is a set of concatenated proplets as the output of the hear mode and used for storage in and retrieval from the agent's on-board database. The other is an equivalent semantic relations graph as the conceptual schema for guiding sequencing in the think-speak mode.

For example, the content of The dog barked. is defined as follows:

11.3.1 FORMAT 1: CONTENT OF The dog barked. AS A SET OF PROPLETS

$$
\begin{bmatrix}
\text{sur:} \\
\text{noun: } \textbf{dog} \\
\text{cat: snp} \\
\text{sem: def sg} \\
\text{fnc: } \textbf{bark} \\
\ldots \\
\text{prn: 14}
\end{bmatrix}
\begin{bmatrix}
\text{sur:} \\
\text{verb: } \textbf{bark} \\
\text{cat: \#n' decl} \\
\text{sem: ind past} \\
\text{arg: } \textbf{dog} \\
\ldots \\
\text{nprn: 14}
\end{bmatrix}
$$

For purposes of storage and retrieval in the agent's content-addressable onboard database (A-memory), the proplets of a content must be order-free. They are connected by a shared prn value, here 14, and the semantic relations of structure, here subject/predicate, shown by the values in bold face.

Navigating from the *dog* to the *bark* proplet is based on the address (bark 14) derived from the *dog* proplet. Navigating from the *bark* proplet back to the *dog* proplet is based on the address (dog 14) derived from the *bark* proplet. This is shown by the following graphical representation of the content:

11.3.2 FORMAT 2: CONTENT OF The dog barked. AS A GRAPH

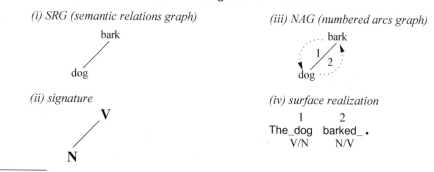

(i) SRG (semantic relations graph)

bark

dog

(ii) signature

V

N

(iii) NAG (numbered arcs graph)

bark

1
2

dog

(iv) surface realization

1   2
The_dog   barked_ .
V/N   N/V

---

[6] Grammatical gender of personal pronouns in indexical and anaphoric use (CLaTR 11) exists also in English, as when calling a ship a she.

The semantic relation of subject/predicate is shown by the ╱ lines in the graphs. There are four views on a content: the (i) SRG (semantic relations graph) connects the core *values* of the proplets; the (ii) signature connects the core *attributes*; the (iii) NAG (numbered arcs graph) supplements the SRG with numbered arcs, which are used in the linear notation of the (iv) surface realization.

Language-dependent surfaces are realized from the *goal proplet* of a traversal. Thus, The dog is realized from the goal proplet of arc 1, and barked_. from the goal proplet of arc 2. Both traversals are along the subject/predicate relation, but arc 1 is in the downward direction ╱ and arc 2 in the upward direction ╱.

While the operations of the hear mode take two proplets as input and produce one or two proplets as output, the navigation rules of the think-speak mode take one input proplet and retrieve one output proplet. Consider the think-speak mode operation **V╱N**, which produces the German surface Der Hund for The dog:

11.3.3 Applying the think-speak operation V╱N

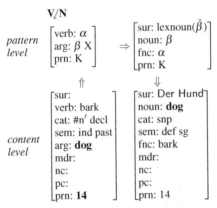

For retrieval of the output, the navigation step uses the address value (dog 14) of the input proplet *bark*. The surface is realized by the lexicalization rule $\text{lexnoun}(\hat{\beta})$, which sits in the sur slot of the goal proplet. It uses the language-dependent variant Hund of the core value dog and the sem values def sg for realizing the German surface Der Hund. In nonlanguage navigation (e.g., activation, reasoning) the lex-rules are switched off.

11.4 Prepositional Phrases

Prepositional phrases consist of a preposition as the functor and a noun as the argument. The semantic kind of the noun is unrestricted in that it may be a concept, e.g., in the water, a name, e.g., in Paris, or an indexical, e.g., in here.

11.4.1 LEXICAL EXAMPLES OF PREPOSITIONS IN GERMAN

sur: auf	sur: über	sur: unter	sur: in	sur: von
noun: n_1	noun: n_1	noun: n_1	noun: n_1	noun: n_1
cat: adnv	cat: adnv	cat: adnv	cat: adnv	cat: adnv
sem: *on*	sem: *above*	sem: *below*	sem: *in*	sem: *of*
mdd:	mdd:	mdd:	mdd:	mdd:
mdr:	mdr:	mdr:	mdr:	mdr:
nc:	nc:	nc:	nc:	nc:
pc:	pc:	pc:	pc:	pc:
prn:	prn:	prn:	prn:	prn:

The core value of a preposition is a substitution variable. Because prepositions like above, below, before, after, etc., are less abstract than the determiner sem values sg, pl, indef, and def, the language-independent counterpart of a preposition is stored as the initial value of the sem slot, using English place holders in *italics*, followed by the determiner values (cf. 11.4.2).

The argument of a preposition may be of unlimited complexity, e.g., in+the_ little_red_house_by_the_lake. Like determiners, prepositions have the core attribute noun, which facilitates the time-linear processing of phrases as in Paris, in the city, in the big old city, in the big old city by the river, etc., with unlimited length.

If a preposition takes a determiner+noun composition (instead of a name or an indexical) as its argument, the time-linear hear mode derivation first combines the preposition and the determiner, e.g., in+the, and then adds the noun, e.g., in+the+garden. The following examples compare the time-linear hear mode derivations of a determiner+noun with a preposition+determiner+noun composition:

11.4.2 DIFFERENT FUNCTION WORD ABSORPTIONS (CLaTR 7.2.5)

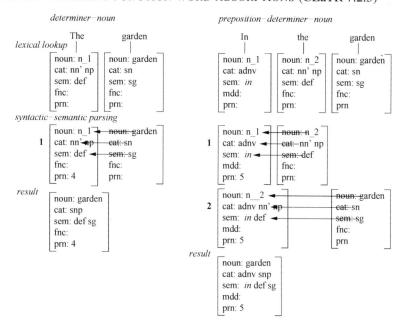

Determiner and preposition proplets are alike in that their core attribute is noun. They differ in that determiners take lexical cat values like sn' snp while the lexical cat value of prepositions is adnv, for adnominal or adverbial modification.

On the left, the determiner+noun derivation (i) substitutes the n_1 value of *the* with the core value of *garden*, (ii) cancels the nn' position with the sn value, (iii) replaces the np value with snp, (iv) adds the sg value to the sem attribute of the former *the* proplet, and (v) discards the *garden* proplet (NLC 13.3.3). The substitution-variable n_1 as the core value of the determiner is used for finding the determiner when it is separated from the noun argument by several modifiers, as in the large, beautiful ... garden.

On the right, the lexical preposition proplet introduces the continuation attribute mdd (modified, for e.g., *sleep*). Step 1 of the time-linear preposition+determiner+noun derivation combines the two lexical function word proplets *in* and *the* into a single noun proplet.[7] Thereby the substitution variable n_1 in the preposition proplet is replaced with the incremented value n_2 of the determiner proplet, the def value of the determiner proplet is added to the preposition's sem slot, and the determiner proplet is discarded. Step 2 fuses the proplet resulting from step 1 with the lexical *garden* proplet: the n_2 substitution variable is replaced by the core value of the *garden* proplet, which is then discarded.

In linear notation, the adverbial use of an elementary adjective, as in Julia slept there, is represented as A|V, while the corresponding construction with a prepositional phrase, as in Julia slept in the garden., is represented as N|V. Graphically, the two constructions differ in the category node of the adverbial:

11.4.3 ELEMENTARY ADVERBIAL VS. PREPOSITIONAL PHRASE

In linear notation, the adnominal use of an elementary modifier is represented as A|N and the phrasal counterpart as N|N (CLaTR 7.3.6; NLC 7.3, 7.4).

11.5 Auxiliaries

There are three kinds of auxiliaries in English, namely do, have, and be, and a larger number of modals, such as can, could, shall, should, will, would, may, might, and must, ought. In the present tense, the auxiliaries have special agreement, i.e., does,

---

[7] As shown in CLaTR 3.5.5, DBS uses the cat value adn (adnominal) for elementary modifiers restricted to nouns, e.g., beautiful, the cat value adv (adverbial) for elementary modifiers restricted

has, and is, while the modals do not.[8] Also, the auxiliaries have a progressive form, e.g., doing, having, and being, while the modals do not.

The auxiliaries do and have have three finite forms do, does, did, and have, has, had, respectively, which are morphologically parallel to the forms of the main verbs and share their pattern of nominative agreement. The auxiliary be has the five finite forms am, is, are, was, and were, which require a special pattern for nominative agreement and may be described schematically as follows:

11.5.1 NOMINATIVE AGREEMENT OF THE AUXILIARY be (FoCL 17.3.1)

				(pnp)	the girls
		(ns3)	he, she	(np13)	we, they
(ns1)	I	(snp)	the boy, John, it	(pro2)	you

[am (ns1' be' v) *]   [is (ns3' be' v) *]   [are (n−s13' be' v) *]
                                              [were (n−s13' be' v) *]

(snp)   the boy, John, it
(ns1)   I

[was (ns13' be' v) *]

Finite forms of the auxiliaries combine with nonfinite forms of the main verbs into complex verb forms. The nonfinite forms are the infinitive, e.g., (to) give, the past participle, e.g., (has) given, and the present participle, e.g., (is) giving.

English infinitives (CLaTR Sect. 15.4) resemble the unmarked present tense form of the main verb, e.g., give. The past participle is marked morphologically in some irregular verbs, e.g., given, but usually coincides with the past tense of the main verb, e.g., worked. The present participle is always marked, as in giving or working.

The infinitive combines with the finite forms of do into the emphatic, e.g., does give or did give. The past participle combines with the finite forms of have into the present perfect, e.g.,, has given or had given. The present participle combines with the finite forms of be into the progressive, e.g., is giving and was giving.

The finite auxiliary forms all have variants with integrated negation, namely don't, doesn't, didn't, haven't, hasn't, hadn't, isn't, aren't, wasn't, and weren't. They have the same combinatorial properties as their unnegated counterparts.

---

to verbs, e.g., beautifully, and the cat value adnv for elementary modifiers which may be applied equally to verbs or nouns, e.g., fast. Because prepositional phrases may always be used adnominally or adverbially, their cat value is adnv as well. Elementary and phrasal adnvs differ in their core attribute, i.e., adj vs. noun.

[8] German auxiliaries and modals have several inflectional forms. For example, the German counterparts to have are habe, hast, hat, haben, habt, and to had are hatte, hattest, hatten, hattet.

The basic categorial structure of combining a finite auxiliary with a nonfinite main verb may be shown schematically as follows:

11.5.2 COMPLEX VERB FORMS OF ENGLISH (FoCL 17.3.2)

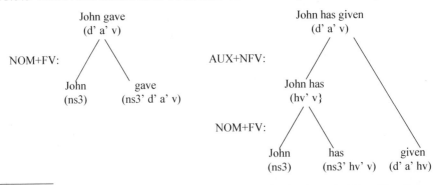

does   give   does give
(ns3' do' v) (d' a' do)   ⟶   (ns3' d' a' v)

has   given   has given
(ns3' hv' v) (d' a' hv)   ⟶   (ns3' d' a' v)

is   giving   is giving
(ns3' be' v) (d' a' be)   ⟶   (ns3' d' a' v)

The nominative agrees with the finite auxiliary, which is why its valency position (here ns3') is located in the category of the auxiliary. The oblique valency positions d' and/or a', in contrast, originate in the nonfinite main verb. That the above auxiliaries are finite is marked lexically by the presence of the v segment in their categories. That the main verb forms are nonfinite is marked lexically by the absence of the v segment. The identity-based agreement between the finite auxiliary and the nonfinite main verb form is expressed in the cat slot of the auxiliary by the segments do (for 'do'), hv (for 'have'), and be (for 'be'), respectively.

The combination of an auxiliary with a nonfinite main verb form, e.g.,, has given, results in a complex verb form which has the same properties in terms of nominative agreement and oblique valency positions as the corresponding finite form of the main verb in question, here gave:

11.5.3 DERIVING BASIC AND COMPLEX VERB FORM (FoCL[9] 17.3.3)

```
        John gave                          John has given
        (d' a' v)                          (d' a' v)

NOM+FV:    /\               AUX+NFV:           /\

    John      gave                    John has
    (ns3)  (ns3' d' a' v)             (hv' v}
                                            /\
                            NOM+FV:        /  \

                                    John      has          given
                                    (ns3)  (ns3' hv' v)  (d' a' hv)
```

---

[9] FoCL is still sign-based and cancels valency positions by deletion instead of #-marking. Compared to the top-down format used in FoCL, the format here is in the now customary time-linear bottom up order.

The two partial derivations end in the same state and may be continued the same.

In English, the auxiliary and its nonfinite main verb take the same adjacent surface positions in main and corresponding subclauses:

11.5.4 ADJACENT POSITION IN ENGLISH MAIN AND SUBCLAUSES

He *had read* the book.
After he *had read* the book, ...

He *did not do* the dishes.
Because he *did not do* the dishes, ...

He *is walking* the dog.
Because he *is walking* the dog, ...

The auxiliaries have take a past participle, do an infinitive, and be a progressive as their nonfinite counterpart.

German, in contrast, has only two auxiliaries, sein and haben, which combine with the past participle of the main verb. A finite auxiliary and a nonfinite transitive verb take different positions in corresponding main and subclauses: in main clauses, the nonfinite verb is in final position ('Distanzstellung'), but in a subordinate clause the nonfinite verb and the auxiliary are adjacent in final position ('Kontaktstellung'):

11.5.5 SEPARATED POSITIONS IN GERMAN MAIN CLAUSES

Er *hat* das Buch *gelesen.*
Nachdem er das Buch *gelesen hat,* ....

Er *ist* zur Schule *gelaufen.*
Weil er zur Schule *gelaufen ist,* ....

Er *soll* die Teller *spülen.*
Weil er die Teller *spülen soll,* ....

'Distanzstellung' in German main clauses is known as 'Satzklammer' (sentence brace). German auxiliaries combine uniformly with the past participle of the main verb, while modals combine with the infinitive, as shown by the third example with *sollen.*

11.6 Subordinating Conjunctions

Examples of subclauses are (i) clausal subjects and objects using, e.g., that, (ii) clausal adnominals with a subject or object gap, using, e.g., who, and (iii) clausal modification using, e.g., when, as their subordinating conjunction. As function words, subordinating conjunctions use a substitution variable as their core value.

11.6.1 LEXICAL SUBORDINATING CONJUNCTIONS

$$
\begin{bmatrix}
\text{sur: that} \\
\text{verb: v\_1} \\
\text{cat:} \\
\text{sem: } that \\
\text{arg:} \\
\text{fnc:} \\
\text{mdr:} \\
\text{nc:} \\
\text{pc:} \\
\text{prn: 14}
\end{bmatrix}
\begin{bmatrix}
\text{sur: who} \\
\text{verb: v\_1} \\
\text{cat:} \\
\text{sem: } who \\
\text{arg: } \emptyset \\
\text{mdd:} \\
\text{mdr:} \\
\text{nc:} \\
\text{pc:} \\
\text{prn: 15}
\end{bmatrix}
\begin{bmatrix}
\text{sur: when} \\
\text{verb: v\_1} \\
\text{cat:} \\
\text{sem: } when \\
\text{arg:} \\
\text{mdd:} \\
\text{mdr:} \\
\text{nc:} \\
\text{pc:} \\
\text{prn: 16}
\end{bmatrix}
$$

The proplets of subordinating conjunctions are special in that they have 10 attributes instead of the standard 9. For example, the additional fnc attribute in the *that* proplet is normally used for connecting an elementary or phrasal subject (11.3.1) or object to the predicate, but needed in subject and object clauses for the same purpose. The mdr attributes are still needed for examples like That John ate the cookie *slowly* surprised Mary.

The following examples have been analyzed in TExer in full declarative detail, which is canonized as the seven to-do's of DBS (11.6.3):

11.6.2 THE SUB-CLAUSE EXAMPLES ANALYZED IN TEXER

1. *clausal subject* (TExer Sect. 2.5)
   That Fido barked amused Mary.
2. *clausal object* (TExer Sect. 2.6)
   Mary heard that Fido barked.
3. *Clausal adnominal modifier with subject gap* (TExer Sect. 3.3)
   The dog which saw Mary barked.
4. *Clausal adnominal modifier with object gap* (TExer Sect. 3.4)
   The dog which Mary saw barked.
5. *Clausal adverbial modification* (TExer Sect. 3.5)
   When Fido barked Mary laughed.

The seven **To-do**s are defined in TExer 1.5.2 as follows:

11.6.3 THE **To-do**'S OF BUILDING A DBS GRAMMAR

1. **<to-do 1>**
   Definition of the content for an example surface
2. **<to-do 2>**
   Graphical hear mode derivation of the content
3. **<to-do 3>**
   Complete sequence of explicit hear mode operation applications

4. **\<to-do 4\>**
   Canonical DBS graph analysis underlying production
5. **\<to-do 5\>**
   List of speak mode operation names with associated surface realizations
6. **\<to-do 6\>**
   Complete sequence of explicit speak mode operation applications
7. **\<to-do 7\>**
   Summary of the system extension and comparison of the hear and speak mode operation applications

English and German are alike in that the grammatical roles of clausal *arguments* as subject, e.g., That Fido barked amused Mary, and as object, e.g., Mary heard that Fido barked, are encoded by word order and the choice of the higher verb. They differ in clausal *adnominals*: English encodes the role as subject, e.g., man who saw Mary, and as object, e.g., man who[10] Mary saw, by word order, but German by means of morphology: der Mann *der* Maria sah (subject) vs. der Mann *den* Maria sah (object). Variation in clausal *modification* is similar in English and German in that it relies on different conjunctions such as when, since, while (temporal), because (reason), where (locational), into (directional), etc.

11.7 Coordinating Conjunctions

The functor-argument relations subject/predicate, object\predicate, and modi-fier|modified are encoded by the values of the noun, fnc, verb, arg, mdr, and mdd attributes. The conjunct–conjunct relations, in contrast, are encoded by the values of the nc (next conjunct) and pc (previous conjunct) attributes. Function words of coor-dination are and, or, but. In the medium of writing, DBS uses the interpunctuation signs ., ?, and ! for extrapropositional conjunction (Ballmer 1978).

Intrapropositionally, conjuncts must be grammatically similar (Bruening and Al Khalaf 2020), while no such constraint holds for extrapropositional coordination: declaratives may follow interrogatives and imperatives, imperatives may follow declar-atives and interrogatives, and interrogatives may follow imperatives and declaratives. Intra- and extrapropositional coordination differ also in that intrapropositional coordi-nation connects conjuncts bidirectionally by cross-copying, while extrapropositional coordination is unidirectional in the direction of time and uses inferencing for occa-sional backward traversal when needed.

In running text, unidirectional extrapropositional forward coordination based on in-terpunctation signs may continue without limit; for a minimal example in complete declarative detail see TExer 2.1.5–2.1.19. For an intrapropositional coordination see TExer Sect. 3.6.

---

[10] With optional use of whom (morphological relic). The word order difference remains.

11.8 Conclusion

In a well-designed software solution, computer scientists distinguish (i) the *declarative specification* and (ii) the *procedural implementation*. The declarative specification presents the conceptual aspect: it must be easily read by humans and at the same time easily implemented in a programming language of choice. This includes the definition of input and output, the functional flow, the abstract data structure, the abstract operation schema, etc., in short the *necessary* properties of the software solution.

A declarative specification may have an open number of procedural implementations which differ in *accidental* properties, i.e., properties inherent in different programming languages and programming styles. A procedural implementation is not only needed practically for using the software solution in applications, but also theoretically as the method of *verifying* the declarative specification.

A topic in computational linguistics well-suited for demonstrating the descriptive power of a declarative specification is the morpho-syntactic mechanisms of syntactic-semantic composition, which natural language controls with a precise mix of (i) function words, (ii) morphology, and (iii) word order. In this chapter, it is demonstrated with detailed declarative specifications of concrete constructions in classical Latin, English, and German.

12. Language vs. Nonlanguage Cognition

A basic distinction in agent-based data-driven DBS is between language and non-language cognition. Language cognition transfers content between agents by means of raw data. Nonlanguage cognition maps between content and raw data inside the focus agent. In language cognition, the speaker's action precedes the hearer's recognition,[1] while in nonlanguage cognition, recognition - including the output of inferencing - precedes action.

Recognition applies a concept type to raw data, resulting in a concept token. In language recognition, the focus agent (hearer) takes raw language-data (surfaces) produced by another agent (speaker) as input, while nonlanguage recognition takes raw nonlanguage-data as input. In either case, the output is a content.

Action adapts a concept type into a token for a purpose. In language action, the focus agent (speaker) produces language-dependent surfaces for another agent (hearer), while nonlanguage action produces intentions for a nonlanguage purpose. In either case, the output is raw data.

In DBS, place holders for concepts make language cognition selfcontained, but provide systematic interaction with nonlanguage cognition. For input-output equivalence between the natural prototype and the artificial reconstruction, the place holders must be properly implemented in nonlanguage cognition.

12.1 Building Blocks and Relations of Cognition

DBS cognition uses the same computational data structure and the same semantic relations of structure for connecting them into content:

12.1.1 COMPUTATIONAL DATA STRUCTURE

The elementary computational data structure of cognition is nonrecursive feature structures with ordered attributes, called proplets.

The relations connecting proplets into content are the classical semantic relations of structure[2]

---

[1] In stored language, e.g. writing, recognition may precede action by thousands of years.

[2] In contradistinction to the semantic relations of the lexicon, i.e. synonymy, antonymy, and hyponymy.

12.1.2 The Semantic Relations of Structure

1. subject/predicate
2. object\predicate
3. modifier|modified
4. conjunct−conjunct

12.2 Example of a Content

In contradistinction to propositions denoting truth values, as in formal logic, propositions *are content* in DBS. A content is defined as a set of proplets, connected by the semantic relations of structure, coded by address. Consider the following example:

12.2.1 Content of Lucy found a big blue square . as a set of proplets

⌈sur: lucy	⌈sur:	⌈sur:	⌈sur:	⌈sur: ⌉
noun: [person x]	**verb: find**	adj: **big**	adj: **blue**	**noun: square**
cat: snp	cat: n′ a′ decl	cat: adn	cat: adnv	cat: sn
sem: nm f	sem: ind past	sem: pad	sem: pad	sem: sg
fnc: find	**arg: [person x] square**	mdd: **square**	mdd:	**fnc: find**
mdr:	mdr:	mdr:	mdr:	mdr: **big**
nc:	nc:	nc: **blue**	nc:	nc:
pc:	pc:	pc:	pc: **big**	pc:
⌊prn: 14	⌊prn: 14	⌊prn: 14	⌊prn: 14	⌊prn: 14 ⌋

The subject/predicate relation is coded between (i) the core feature **[noun: [person x]]** of lucy and the continuation feature **[arg: [person x]** *square*] of find, and (ii) the core feature **[verb: find]** of find and the continuation feature **[fnc: find]** of lucy. Similarly, the object\predicate relation is coded between (i) the core feature **[noun: square]** of square and the continuation feature **[arg:** [person x] **square]** of find and (ii) the core feature of **[verb: find]** of find and the continuation feature **[fnc: find]** of square (bidirectional.)

12.3 Content as Input to the Speak Mode

The speak mode takes a content like 12.2.1 as input (i, ii), activates it by time-linear navigation along the semantic relations of structure (iii), and results in a language-dependent surface as output (iv), realized as raw data in a medium of choice:

12.3.1 Semantic Relations Underlying Speak Mode Derivation

(i) SRG (semantic relations graph)   (iii) NAG (numbered arcs graph)

(ii) signature

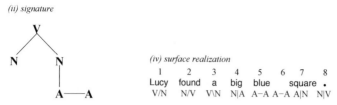

(iv) surface realization

1	2	3	4	5	6	7	8
Lucy	found	a	big	blue	square		.
V/N	N/V	V\N	N|A	A–A	A–A	A|N	N|V

The / line is the subject/predicate relation no matter whether it is long, as in the graphs (i), (ii), (iii), or short and of a somewhat different angle (for better formatting in print) in the (iv) surface realization, and similarly for the other connectives.

12.4 Content as Output of the Hear Mode

The speaker's output is the hearer's input. The hearer reconstructs the speaker's input content by means of a time-linear surface-compositional derivation:

12.4.1 HEAR MODE DERIVATION OF 12.3.1, *(iv)*

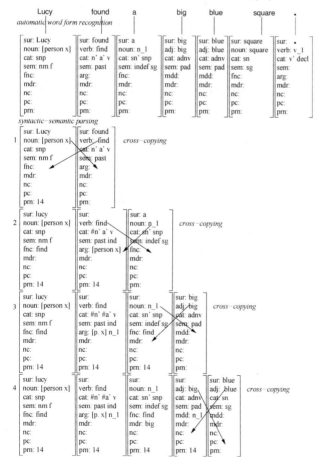

```
5  ⎡sur: lucy        ⎤⎡sur:            ⎤⎡sur:            ⎤⎡sur:        ⎤⎡sur:        ⎤⎡sur: square   ⎤  absorption
   ⎢noun: [person x] ⎥⎢verb: find      ⎥⎢noun: n_1       ⎥⎢adj: big    ⎥⎢adj: blue   ⎥⎢noun: square  ⎥  with
   ⎢cat: snp         ⎥⎢cat: #n' #a' v  ⎥⎢cat: sn' snp    ⎥⎢cat: adnv   ⎥⎢cat: adnv   ⎥⎢cat: sn       ⎥  simultaneous
   ⎢sem: nm f        ⎥⎢sem: past ind   ⎥⎢sem: indef sg   ⎥⎢sem: pad    ⎥⎢sem: pad    ⎥⎢sem: sg       ⎥  substitution
   ⎢fnc: find        ⎥⎢arg: [p. x] n_1 ⎥⎢fnc: find       ⎥⎢mdd: n_1    ⎥⎢mdd:        ⎥⎢fnc:          ⎥
   ⎢mdr:             ⎥⎢mdr:            ⎥⎢mdr: big        ⎥⎢mdr:        ⎥⎢mdr:        ⎥⎢mdr:          ⎥
   ⎢nc:              ⎥⎢nc:             ⎥⎢nc:             ⎥⎢nc: blue    ⎥⎢nc:         ⎥⎢nc:           ⎥
   ⎢pc:              ⎥⎢pc:             ⎥⎢pc:             ⎥⎢pc:         ⎥⎢pc: big     ⎥⎢pc:           ⎥
   ⎣prm: 14          ⎦⎣prm: 14         ⎦⎣prm: 14         ⎦⎣prm: 14     ⎦⎣prm: 14     ⎦⎣prm:          ⎦

6  ⎡sur: lucy        ⎤⎡sur:              ⎤⎡sur:           ⎤⎡sur:        ⎤⎡sur:        ⎤⎡sur: •        ⎤  absorption
   ⎢noun: [person x] ⎥⎢verb: find        ⎥⎢noun: square   ⎥⎢adj: big    ⎥⎢adj: blue   ⎥⎢verb: v_1     ⎥
   ⎢cat: snp         ⎥⎢cat: #n' #a' v    ⎥⎢cat: sn' snp   ⎥⎢cat: adnv   ⎥⎢cat: adnv   ⎥⎢cat: v' decl  ⎥
   ⎢sem: nm f        ⎥⎢sem: past ind     ⎥⎢sem: indef sg  ⎥⎢sem: pad    ⎥⎢sem: pad    ⎥⎢sem:          ⎥
   ⎢fnc: find        ⎥⎢arg: [p. x] square⎥⎢fnc: find      ⎥⎢mdd: square ⎥⎢mdd:        ⎥⎢arg:          ⎥
   ⎢mdr:             ⎥⎢mdr:              ⎥⎢mdr: big       ⎥⎢mdr:        ⎥⎢mdr:        ⎥⎢mdr:          ⎥
   ⎢nc:              ⎥⎢nc:               ⎥⎢nc:            ⎥⎢nc: blue    ⎥⎢nc:         ⎥⎢nc:           ⎥
   ⎢pc:              ⎥⎢pc:               ⎥⎢pc:            ⎥⎢pc:         ⎥⎢pc: big     ⎥⎢pc:           ⎥
   ⎣prm: 14          ⎦⎣prm: 14           ⎦⎣prm: 14        ⎦⎣prm: 14     ⎦⎣prm: 14     ⎦⎣prm:          ⎦
result
   ⎡sur: lucy        ⎤⎡sur:                ⎤⎡sur:           ⎤⎡sur:        ⎤⎡sur:        ⎤
   ⎢noun: [person x] ⎥⎢verb: find          ⎥⎢noun: square   ⎥⎢adj: big    ⎥⎢adj: blue   ⎥
   ⎢cat: snp         ⎥⎢cat: #n' #a' decl   ⎥⎢cat: sn' snp   ⎥⎢cat: adnv   ⎥⎢cat: adnv   ⎥
   ⎢sem: nm f        ⎥⎢sem: past ind       ⎥⎢sem: indef sg  ⎥⎢sem: pad    ⎥⎢sem: pad    ⎥
   ⎢fnc: find        ⎥⎢arg: [p. x] square  ⎥⎢fnc: find      ⎥⎢mdd: square ⎥⎢mdd:        ⎥
   ⎢mdr:             ⎥⎢mdr:                ⎥⎢mdr: big       ⎥⎢mdr:        ⎥⎢mdr:        ⎥
   ⎢nc:              ⎥⎢nc:                 ⎥⎢nc:            ⎥⎢nc: blue    ⎥⎢nc:         ⎥
   ⎢pc:              ⎥⎢pc:                 ⎥⎢pc:            ⎥⎢pc:         ⎥⎢pc: big     ⎥
   ⎣prm: 14          ⎦⎣prm: 14             ⎦⎣prm: 14        ⎦⎣prm: 14     ⎦⎣prm: 14     ⎦
```

As required for successful natural language communication, the output content of this hear mode derivation equals the input content 14.9.2 to the speak mode.

A DBS hear mode derivation is (a) *surface-compositional* because each lexical item has a concrete sur value and there are no surfaces without a proplet analysis. The derivation order is (b) *time-linear*, as shown by the stair-like addition of a next word proplet (incremental loading). The application of operations is (c) *data-driven* by the incoming sequence of word form proplets provided by automatic word form recognition.

12.5 Nonlanguage Cognition Provides Place Holder Values

In DBS, content words such as square, blue, and find use *concepts* as core and continuation values. Indexicals like the we, you, they, here, and now use *pointing* at STAR values of the agent's onboard orientation system (OBOS). Function words like determiners, prepositions, and conjunctions use *substitution variables*.

The following noun proplets have the same attributes, but differ in values:

12.5.1 Some Lexical Content Word Proplets: Nouns

```
⎡sur: square  ⎤⎡sur: squares ⎤⎡sur: John         ⎤⎡sur: Mary         ⎤⎡sur: Gorch Fock⎤
⎢noun: square ⎥⎢noun: square ⎥⎢noun: [person 10] ⎥⎢noun: [person 11] ⎥⎢noun: [bark 12]⎥
⎢cat: sn      ⎥⎢cat: pn      ⎥⎢cat: snp          ⎥⎢cat: snp          ⎥⎢cat: snp       ⎥
⎢sem: sg      ⎥⎢sem: pl      ⎥⎢sem: sg m         ⎥⎢sem: sg f         ⎥⎢sem: sg f      ⎥
⎢fnc:         ⎥⎢fnc:         ⎥⎢fnc:              ⎥⎢fnc:              ⎥⎢fnc:           ⎥
⎢mdr:         ⎥⎢mdr:         ⎥⎢mdr:              ⎥⎢mdr:              ⎥⎢mdr:           ⎥
⎢nc:          ⎥⎢nc:          ⎥⎢nc:               ⎥⎢nc:               ⎥⎢nc:            ⎥
⎢pc:          ⎥⎢pc:          ⎥⎢pc:               ⎥⎢pc:               ⎥⎢pc:            ⎥
⎣prn:         ⎦⎣prn:         ⎦⎣prn:              ⎦⎣prn:              ⎦⎣prn:           ⎦
```

The common noun proplets square and squares differ in grammatical number (cat slot), needed for agreement with the finite verb in the 3rd person singular indicative present. The name proplets John and Mary differ in grammatical gender, needed for coreference with the correct personal pronoun, e.g., he vs. she. The name proplet Gorch Fock illustrates the naming of inanimate objects like ships, mountains, and cities. Name proplets occur also in the plural[3], as in the Millers, which have the cat value pnp.

Function words with the same attributes as nouns are the determiners. They differ from content words in that their core value is a substitution variable, here n_1:

12.5.2 SOME LEXICAL FUNCTION WORD PROPLETS: DETERMINERS

$$
\begin{bmatrix} \text{sur: a(n)} \\ \text{noun: n\_1} \\ \text{cat: sn' snp} \\ \text{sem: indef sg} \\ \text{fnc:} \\ \ldots \\ \text{prn:} \end{bmatrix}
\begin{bmatrix} \text{sur: some} \\ \text{noun: n\_1} \\ \text{cat: nn' np} \\ \text{sem: indef sel} \\ \text{fnc:} \\ \ldots \\ \text{prn:} \end{bmatrix}
\begin{bmatrix} \text{sur: all} \\ \text{noun: n\_1} \\ \text{cat: pn' pnp} \\ \text{sem: pl exh} \\ \text{fnc:} \\ \ldots \\ \text{prn:} \end{bmatrix}
\begin{bmatrix} \text{sur: every} \\ \text{noun: n\_1} \\ \text{cat: sn' snp} \\ \text{sem: pl exh} \\ \text{fnc:} \\ \ldots \\ \text{prn:} \end{bmatrix}
\begin{bmatrix} \text{sur: the} \\ \text{noun: n\_1} \\ \text{cat: nn' np} \\ \text{sem: def} \\ \text{fnc:} \\ \ldots \\ \text{prn:} \end{bmatrix}
$$

For grammatical agreement, certain grammatical values of determiners equal those of common nouns, e.g., sg and pl. Additional grammatical values like def, indef, sn', pn', nn', sel, and exh characterize grammatical properties needed for determiner∪noun combination by absorption (12.6.1).

12.6 Function Word Absorbs Content Word

The hear mode operation combining a determiner with a common noun takes two input proplets and produces one output proplet (function word absorption):

12.6.1 The ABSORBING square WITH DET∪CN

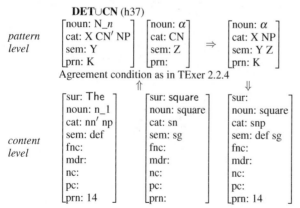

The input to the operation are the proplets *the* and *square*. The content word is absorbed into the function word by replacing the substitution variable n_1 with the place holder value square and adjusting the cat and sem values.

---

[3] Pace Russell's (1905) "uniqueness condition" for "proper" names.

Thereby all instances of the substitution variable accumulated so far are replaced by the substitution value ('simultaneous' substitution, 12.4.1, line 5). It follows from the time-linear derivation order that once a substitution variable has been replaced by a place holder, it cannot be used again (functional universal). However, subsequent reference to, e.g., the big blue square is enabled by an adnominal modifier clause, e.g., which ..., or a pronoun, e.g., it in anaphoric use (CLaTR 11).

12.7 Type-Token Matching in Recognition and Action

The implementation of concepts like square or blue is language-independent work in cognitive psychology and robotics, while the construction of syntactic-semantic DBS grammars in the speak and the hear mode is language-dependent work in linguistics. Separating the empirical work on implementing language-independent concepts vs. language-dependent DBS grammars is made possible by the use of place holder values, which have abstract definitions based on natural science:

12.7.1 CONCEPT TYPE square AS DEFINED IN SCIENCE (GEOMETRY)

$$
\begin{bmatrix}
\text{edge 1: } \alpha \text{ cm} \\
\text{angle 1/2: } 90^\circ \\
\text{edge 2: } \alpha \text{ cm} \\
\text{angle 2/3: } 90^\circ \\
\text{edge 3: } \alpha \text{ cm} \\
\text{angle 3/4: } 90^\circ \\
\text{edge 4: } \alpha \text{ cm} \\
\text{angle 4/1: } 90^\circ
\end{bmatrix}
$$

The functioning of this nonlanguage recognition concept may be shown as follows:

12.7.2 RECOGNITION: RAW DATA MATCHING TYPE RESULT IN TOKEN

concept type
$$
\begin{bmatrix}
\text{edge 1: } \alpha \text{ cm} \\
\text{angle 1/2: } 90^\circ \\
\text{edge 2: } \alpha \text{ cm} \\
\text{angle 2/3: } 90^\circ \\
\text{edge 3: } \alpha \text{ cm} \\
\text{angle 3/4: } 90^\circ \\
\text{edge 4: } \alpha \text{ cm} \\
\text{angle 4/1: } 90^\circ
\end{bmatrix}
\Rightarrow
\begin{bmatrix}
\text{edge 1: 2 cm} \\
\text{angle 1/2: } 90^\circ \\
\text{edge 2: 2 cm} \\
\text{angle 2/3: } 90^\circ \\
\text{edge 3: 2 cm} \\
\text{angle 3/4: } 90^\circ \\
\text{edge 4: 2 cm} \\
\text{angle 4/1: } 90^\circ
\end{bmatrix}
$$
concept token

⇑ matching

raw data
edge 1: 2 cm
angle 1/2: 90°
edge 2: 2 cm
angle 2/3: 90°
edge 3: 2 cm
angle 3/4: 90°
edge 4: 2 cm
angle 4/1: 90°

The type defines the concept of a *square*. Replacing its variables with constants, here α with *2cm*, results in a concept token. The constants are measurements of raw input data provided by the agent's interface component.

More schematically, the agent's nonlanguage recognition of a square by means of type-token matching is as follows:

12.7.3 USING A CONCEPT TYPE IN NONLANGUAGE RECOGNITION

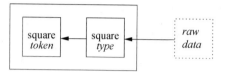

The raw input data are provided by the agent's interface component. They are recognized as an instance of the two-dimensional shape square because the agent's vision sensor detects four lines of equal length and an intersections angle of 90°.

The action counterpart to 12.7.2 is the agent's cognition adapting a concept type into a concept token for a purpose:

12.7.4 ACTION: TYPE-TOKEN ADAPTATION RESULTS IN RAW DATA

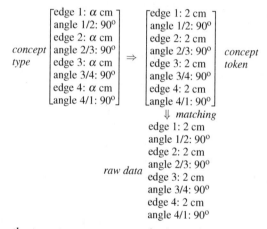

Adapting the type to a purpose results in a token which is realized as raw data.

More schematically, the agent's nonlanguage action of drawing a square by means of a type-token adaptation is as follows:

12.7.5 USING A CONCEPT TOKEN IN NONLANGUAGE ACTION

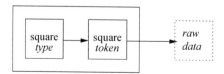

The definition of concept types, corresponding concept tokens, and raw data relies on the natural sciences, here geometry. Type-token matching in recognition and action is an instance of computational pattern matching in DBS (CC 6).

Once an element of a semantic field, e.g., square, has been analyzed as a type, other elements of the class may be treated more or less routinely in a similar way:

12.7.6 ELEMENTS IN THE CLASS OF TWO-DIMENSIONAL OBJECTS

$$
\begin{bmatrix}
\text{place holder: equilateral triangle} \\
\text{sensory modality: vision} \\
\text{semantic field: two-dim geo.} \\
\text{content kind: concept} \\
\text{shape:} \begin{bmatrix} \text{edge 1: } \alpha \text{ cm} \\ \text{angle 1/2: } 60^\circ \\ \text{edge 2: } \alpha \text{ cm} \\ \text{angle 2/3: } 60^\circ \\ \text{edge 3: } \alpha \text{ cm} \\ \text{angle 3/4: } 60^\circ \\ \text{samples: a, b, c,...} \end{bmatrix} \\
\text{samples: a, b, c,...}
\end{bmatrix}
\quad
\begin{bmatrix}
\text{place holder: rectangle} \\
\text{sensory modality: vision} \\
\text{semantic field: two-dim geo.} \\
\text{content kind: concept} \\
\text{shape:} \begin{bmatrix} \text{edge 1: } \alpha \text{ cm} \\ \text{angle 1/2: } 90^\circ \\ \text{edge 2: } \beta \text{ cm} \\ \text{angle 2/3: } 90^\circ \\ \text{edge 3: } \alpha \text{ cm} \\ \text{angle 3/4: } 90^\circ \\ \text{edge 4: } \beta \text{ cm} \\ \text{angle 4/1: } 90^\circ \end{bmatrix} \\
\text{samples: a', b', c',...}
\end{bmatrix}
\quad
\begin{bmatrix}
\text{place holder: square} \\
\text{sensory modality: vision} \\
\text{semantic field: two-dim geo.} \\
\text{content kind: concept} \\
\text{shape:} \begin{bmatrix} \text{edge 1: } \alpha \text{ cm} \\ \text{angle 1/2: } 90^\circ \\ \text{edge 2: } \alpha \text{ cm} \\ \text{angle 2/3: } 90^\circ \\ \text{edge 3: } \alpha \text{ cm} \\ \text{angle 3/4: } 90^\circ \\ \text{edge 4: } \alpha \text{ cm} \\ \text{angle 4/1: } 90^\circ \end{bmatrix} \\
\text{samples: a'', b'', c'',...}
\end{bmatrix}
$$

For retrieving the correct type, i.e., the one best matching the raw data at hand, concept analyses are embedded into feature structures which specify the sensory modality, the semantic field, and whatever else is useful to aid retrieval of the type most suitable for matching the raw data.

The method of type-token matching in recognition and action illustrated above has been extended to the recognition and production of colors in CC 11.3.2, 11.3.4, and 11.3.5. In science, the work of OCR (optical character recognition) and ASR (automatic speech recognition) is largely based on statistics, but the method of type-token matching may be superimposed. In industrial applications, the electronic tongue (Winquist 2008) and the electronic nose (Persaud and George 1982) model the natural prototype using natural science. Combining these modalities in the nonlanguage cognition of a DBS robot may provide many place holder values with explicit operational counterparts.

12.8 Language Communication

The speak and the hear mode of language cognition reuse the mechanisms of nonlanguage recognition and action for new functions. Reuse of earlier mechanisms in the evolutionary transition from nonlanguage to language cognition is in the spirit of Charles Darwin, but cannot be accounted for in systems which use a sign-based substitution-driven ontology.

Nonlanguage and language cognition are alike in that they apply type-token matching to raw data input. They differ in that nonlanguage cognition applies type-token matching to nonlanguage content, while language cognition applies it to language surfaces. In the medium of speech, a surface token differs from its type by specifying volume, pitch, speed, timbre, etc., and in the medium of writing by specifying font, size, color, etc., i.e., what Aristotle would call the accidental properties.

The extension of type-token matching from nonlanguage cognition to language cognition may shown schematically as follows:

12.8.1 THE EVOLUTION OF LANGUAGE

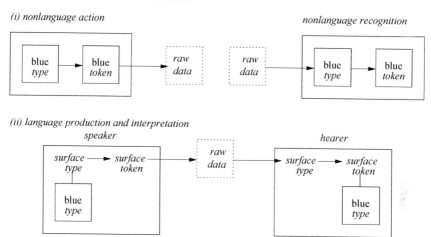

(i) nonlanguage action

nonlanguage recognition

(ii) language production and interpretation

Because the transfer of content is based on raw data, (a) the concept types, (b) the language dependent surface types, and (c) the conventions connecting (a) and (b) exist solely in the respective cognitions of speaker and hearer (anything else would be reification). It presupposes that speaker and hearer have learned the same natural language. In addition, the speaker must be able to produce surface types as tokens and the hearer must be able to recognize the surface token by means of a matching type.

Type-token adaptation in speak mode surface production may be illustrated as follows (medium of writing):

12.8.2 SPEAK MODE: FROM CONTENT TO SURFACE TYPE TO RAW DATA

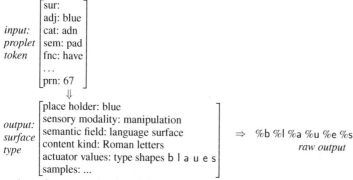

The input, i.e., the content token *blue* of nonlanguage cognition, retrieves the corresponding language-dependent surface, here the type of German b l a u e s, based on a list which provides allomorphs using the input proplet's core, cat, and sem values.

This output serves as input to a realization operation which adapts the surface type into a token, realized as raw data.

Type-token instantiation in hear mode surface recognition may be illustrated as follow

12.8.3 HEAR MODE: RAW DATA TO SURFACE TYPE TO CONTENT

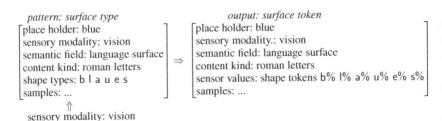

The input consists of raw data which are provided by the agent's vision sensors and matched by the letters' shape types provided by the agent's memory. The output replaces the shape types, here b l a u e s, with the matching raw data resulting in shape tokens; shown as b% l% a% u% e% s%, which record the accidental properties. The function crucial for the understanding of the hearer, however, is using the place holder, here *blue*, for the lexical look-up of the correct nonlanguage concept.

12.9 Conclusion

On the one hand, the lines and angles of two-dimensional geometry (12.7) have counterparts in neurology, such as the line, edge, and angle detectors in the optical cortex of the cat (Hubel and Wiesel 1962), and the iconic or sensory memories from which the internal image representations are built (Sperling 1960) and Neisser (1967). On the other hand, robotic vision (Wiriyathammabhum et al. 2016) applies the natural science of optics in ways which differ from the natural prototype (Pylyshyn 2009).

This is analogous to the difference between the natural flight of (i) birds, bats, and butterflies (flapping wings), and the artificial flight of (ii) air planes (fixed wings), and (iii) helicopters (rotors), all of which satisfy the laws of aerodynamics (CLaTR 11). The list goes on with differences in earth-bound locomotion (legs vs. wheels), and power supply (metabolism vs. electricity).

Input-output equivalence with the natural prototype is not in conflict with alternative (artificial) processing methods for the place holder values. As illustrated in 12.4.1, input-output equivalence affects macro-processing, while alternative uses of the natural sciences affect micro-processing. This mutual independence/interaction between language and nonlanguage cognition is based on the largely language-independent place holder values in language cognition.

13. Grammatical Disambiguation

By combining concatenation operations of constant complexity with a strictly time-linear derivation order, the computational complexity degree of DBS is linear time (TCS'92). The only way to increase DBS complexity above linear would be a recursive ambiguity in the hear mode. In natural language, however, recursive ambiguity is prevented by grammatical disambiguation.[1]

An example of grammatically disambiguating a nonrecursive ambiguity is the 'garden path' sentence The horse raced by the barn fell (Bever 1970). The continuation horse+raced introduces a local ambiguity between *horse raced* (active) and *horse which was raced* (passive), leading to two parallel derivation strands up to and including barn. Depending on continuing after barn with an interpunctuation or a verb, one of the [-global] readings (FoCL 11.3) is grammatically eliminated.

An example of grammatically disambiguating a recursive ambiguity is The man who loves the woman who loves Tom who Lucy loves, with the subordinating conjunction who. Depending on whether the continuation after who is a verb or a noun, one of the two [-global] readings is grammatically eliminated (momentary choice between who being subject or object).

13.1 Degrees of Computational Complexity

Given an algorithm taking an input of length n (n>1), its time complexity is commonly estimated (i) by counting the number of primitive operations needed for adding a next input item and (ii) the increase in the number of operations with the increase of the length n. The basic complexity degrees are a linear, polynomial, exponential, or unbounded increase with the length of the input:

13.1.1 BASIC DEGREES OF COMPLEXITY

1. *Linear complexity*: $1 \cdot n$, $2 \cdot n$, $3 \cdot n$, $4 \cdot n$, etc. (e.g., 2, 4, 6, 8, ... for $n=2$)
2. *Polynomial complexity*: n^1, n^2, n^3, n^4, etc. (e.g., 2, 4, 9, 16, 25, 36, 49, 64, ... for $n=2$))
3. *Exponential complexity*: 1^n, 2^n, 3^n, 4^n, etc. (e.g., 1, 4, 8, 16, 32, 64, 128, 256, ... for $n=2$))
4. *Undecidable*: $n \cdot \infty$

In praxi, the most important distinction is between the computationally *tractable* and *intractable* complexity degrees. As shown by Garey and Johnson (1979), the boundary is between the (2) polynomial and the (3) exponential algorithms:

13.1.2 TIMING OF POLYNOMIAL VS. EXPONENTIAL ALGORITHMS

time complexity	problem size n		
	10	50	100
n^3	0.001 seconds	0.125 seconds	1.0 seconds
2^n	0.001 seconds	35.7 years	10^{15} centuries

The *primitive operation*[2] used is adding the next word (i.e., the minimum). The respective application numbers are shown for lengths 10, 50, and 100.

13.2 Orthogonal LAG and PSG Complexity Hierarchies

Two complexity hierarchies are orthogonal if they classify certain formal languages differently. For example, in PSG the formal languages $a^n b^n$ and WW^r are in the same complexity class, polynomial, but in different classes, C1 (linear) vs. C2 (polynomial), in LAG. In PSG, $a^n b^n$ and $a^n b^n c^n$ are in different complexity classes, polynomial vs. exponential, but in the same class, C1 (linear), in LAG. The formal language L_{no} is polynomial in PSG, but exponential in LAG. The reason for these differences is (i) the substitution-driven derivation of PSG and (ii) the data-driven derivation of LAG.

Substitution-driven PSG favors input which is *pairwise inverse*, like abcd dcba. Formal languages which require no more than this correspondence, are called context-free and of polynomial complexity, but formal languages which exceed the pairwise inverse correspondence are computationally intractable in PSG.[3]

Data-driven LAG favors input which is *not recursively ambiguous*. Unambiguous languages, such as $a^n b^n$, $a^n b^n c^n$, $a^n b^n c^n d^n$, etc., a^{2^n}, $a^{n!}$, and single return languages, such as WW^r, WW, and WWW, are computationally tractable, but languages which

---

[2] Earley (1970) characterizes a primitive operation as "in some sense the most complex operation performed by the algorithm whose complexity is independent of the size of the grammar and the input string." The nature of the primitive operation varies from one grammar formalism to the next.

   For example, Earley chose the operation of *adding a state to a state set* as the primitive operation of his famous algorithm for context-free grammars (FoCL 9.3). In LA Grammar, the subclass of C-LAGs uses a *rule application* as its primitive operation.

[3] According to Harrison (1978, p.219f.) and Ginsburg (1980, p.8), it is 'doubtful' that the structure of context-free PSG approximates the syntax of the programming languages. In other words, the programming languages must be computationally tractable but certainly not pairwise inverse.

are recursively ambiguous with a degree greater 2, such as 3SAT, SUBSET-SUM, L_{no}, and HCFL (Greibach 1973) are computationally intractable in LAG/DBS,
The two complexity hierarchies may be compared graphically as follows:

13.2.1 ORTHOGONAL RELATION BETWEEN C AND CF LANGUAGES

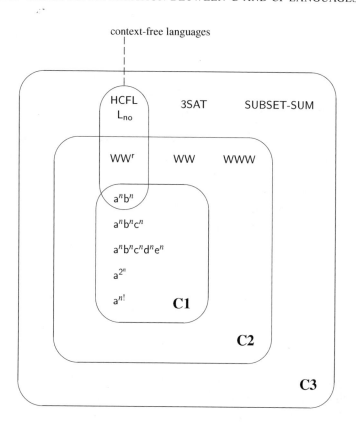

That LAG classifies L_{no} with inherently complex 3SAT and SUBSET-SUM is not be-cause L_{no} is particularly complex, but because it is recursively ambiguous. That PSG classifies $a^n b^n$ and $a^n b^n c^n$ in different classes, context-free (polynomial) vs. context-sensitive (exponential), is not because $a^n b^n c^n$ is particularly complex as compared to $a^n b^n$, but because it is not pairwise. That PSG classifies WW^r and WW in different classes, context-free vs. context-sensitive, is not because WW is particularly complex as compared to WW^r, but because it is not inverse.

13.3 Comparing Explicitly Defined Examples in PSG and DBS

Formal languages critical for distinguishing the complexity hierarchies of PSG and LAG are $a^n b^n$ and $a^n b^n c^n$:

13.3.1 EXPLICIT PSGS FOR THE FORMAL LANGUAGES $a^n b^n$ AND $a^n b^n c^n$

$a^n b^n$ *(polynomial n^3)*
$V =_{def} \{S, a, b\}$
$V_T =_{def} \{a, b\}$
$P =_{def} \{S \to a\,S\,B$
$\phantom{P =_{def} \{}S \to a\,b\}$

$a^n b^n c^n$ *(exponential)*
$V =_{def} \{S, B, C, D_1, D_2, a, b, c\}$
$V_T =_{def} \{a, b, c\}$

$P =_{def} \{S \to a\,S\,B\,C,$	*rule 1*
$S \to a\,b\,C,$	*rule 2*
$C\,B \to D_1\,B$	*rule 3a*
$D_1\,B \to D_1\,B$	*rule 3b*
$D_1\,D_2 \to B\,D_2$	*rule 3c*
$B\,D_2 \to B\,C$	*rule 3d*
$B\,b \to b\,b$	*rule 4*
$B\,C \to B\,c$	*rule 5*
$c\,C \to c\,c\}$	*rule 6*

The PSG for $a^n b^n c^n$ generates pairwise inverse aSBC, aaSBCBC, aaaSBCBCBC, etc. with rule 1 and concludes with rule 2. Then the BCBCBC... sequence is changed into lower case and reordered step by step into bbb...ccc... with rules 3a–6. The rules compute *possible substitutions* and distinguish between nonterminal (e.g., B) and terminal (e.g., b) symbols.

The operations of a LAG, in contrast, compute *possible continuations* and distinguish between the surface and the category of an input, e.g., [aaabb (abb)] (FoCL 10.4.1), running the derivation via the category, used as counter.

13.3.2 EXPLICIT LAGS FOR $a^n b^n$ AND $a^n b^n c^n$

$a^n b^n$ *(linear)*
$LX =_{def} \{[a\,(a)], [b\,(b)]\}$
$ST_S =_{def} \{[(a)\,\{r_1, r_2\}]\}$
$r_1 : (X)\,(a) \Rightarrow (aX)\,\{r_1, r_2\}$
$r_2 : (aX)\,(b) \Rightarrow (X)\,\{r_2\}$
$ST_F =_{def} \{[\varepsilon\,\text{rp}_2]\}.$

$a^n b^n c^n$ *(linear)*
$LX =_{def} \{[a\,(a)], [b\,(b)], [c\,(c)]\}$
$ST_s =_{def} \{[(a)\,\{r_1, r_2\}]\}$
$r_1 : (X)\,(a) \Rightarrow (aX)\,\{r_1, r_2\}$
$r_2 : (aX)\,(b) \Rightarrow (Xb)\,\{r_2, r_3\}$
$r_3 : (bX)\,(c) \Rightarrow (X)\,\{r_3\}$
$ST_F =_{def} \{[\varepsilon\,\text{rp}_3]\}.$

Another language pair critical for the distinction between PSG and LAG are inverse WW^r and repeating WW. Both are pairwise, but inverse WW^r is context-free (n^3 polynomial), while repeating WW is context-sensitive (exponential) in PSG:

13.3.3 EXPLICIT PSG FOR WW^r AND INFORMAL FOR WW

WW^r *(polynomial n^3)*
$V =_{def} \{S, a, b, c, d\}$
$V_T =_{def} \{a, b, c, d\}$
$P =_{def} \{S \to a\,S\,a,$
$S \to b\,S\,b,$
$S \to c\,S\,c,$
$S \to d\,S\,d,$
$S \to a\,a,$
$S \to b\,b,$
$S \to c\,c,$
$S \to d\,d\}$

WW *(exponential)*
Similar to the PSG for $a^n b^n c^n$ (13.3.1), the derivation generates intermediate expressions like aSA, abSBA, abcSCBA, etc., and then reorders in lower case.

In LAG, WW^r and WW are in the same complexity class, namely n^2 polynomial:

13.3.4 EXPLICIT LAGS FOR WW^r AND WW

WW^r (*polynomial* n^2)
LX $=_{def}$ {[a (a)], [b (b)], [c (c)], [d (d)] ... }
ST$_S$ $=_{def}$ {[(seg$_c$) {r$_1$, r$_2$}]}, where seg$_c$ ε {a, b, c, d, ... }
r$_1$: (X) (seg$_c$) $\Rightarrow$ (seg$_c$ X) {r$_1$, r$_2$}
r$_2$: (seg$_c$ X) (seg$_c$) $\Rightarrow$ (X) { r$_2$ }
ST$_F$ $=_{def}$ {[ε rp$_2$] }

WW (*polynomial* n^2)
LX $=_{def}$ {[a (a)], [b (b)], [c (c)], [d (d)] ... }
ST$_S$ $=_{def}$ {[(seg$_c$) {r$_1$, r$_2$}]}, where seg$_c$ ε {a, b, c, d, ... }
r$_1$: (X)   (seg$_c$) $\Rightarrow$ (X seg$_c$) {r$_1$, r$_2$}
r$_2$: (seg$_c$ X) (seg$_c$) $\Rightarrow$ (X)   { r$_2$}
ST$_F$ $=_{def}$ {[ε rp$_2$]}

In WW^r, rule r$_1$ adds the counterpart letter (seg$_c$) at the beginning of the category, but at the end in WW.

The recursive ambiguity of WW^r and WW is of the kind *single return* (FoCL 11.5.3): in each derivation step, the rule package {r$_1$, r$_2$} of r$_1$ calls two input-compatible rules (ambiguity), but the continuation split is disambiguated by the following input.

We complete the orthogonal relation between the PSG and the LAG complexity hierarchies with the noise language L$_{no}$, which is context-free (polynomial) in PSG, but C3 (exponential) in LAG. Devised by D. Applegate, its expressions consist of an arbitrary sequence of 0 and 1, followed by the separation symbol #, followed by an inverse copy with arbitrarily missing symbols of the initial sequence. Thus, when the initial sequence is read in by a LAG, it is not known until the end which pre-separation digits turn out to be genuine and which are noise:

13.3.5 EXPLICIT PSG AND LAG FOR L$_{no}$

L$_{no}$ in PSG (context-free, polynomial)

S $\rightarrow$ 1S1
S $\rightarrow$ 1S
S $\rightarrow$ 0S0
S $\rightarrow$ 0S
S $\rightarrow$ #

L$_{no}$ in LAG (C3, exponential)

LX $=_{def}$ {[0 (0)], [1 (1)], [# (#)]}
ST$_S$ $=_{def}$ {[(seg$_c$) {r$_1$, r$_2$, r$_3$, r$_4$, r$_5$}] },
        where seg ε {0, 1}.
r$_1$: (seg$_c$)(seg$_d$) $\Rightarrow$ ε {r$_1$, r$_2$, r$_3$, r$_4$, r$_5$}
r$_2$: (seg$_c$)(seg$_d$) $\Rightarrow$ (seg$_d$){r$_1$, r$_2$, r$_3$, r$_4$, r$_5$}
r$_3$: (X)(seg$_c$) $\Rightarrow$ (X){r$_1$ r$_2$, r$_3$, r$_4$, r$_5$}
r$_4$: (X)(seg$_c$) $\Rightarrow$ (seg$_c$ X){r$_1$ r$_2$, r$_3$, r$_4$, r$_5$}
r$_5$: (X)(#) $\Rightarrow$ (X){r$_6$}
r$_6$: (seg$_c$ X)(seg$_c$) $\Rightarrow$ (X) {r$_6$}
ST$_F$ $=_{def}$ {[ε rp$_6$]}

The complexity hierarchies of PSG and LAG may be summarized as follows:

13.3.6 Complexity Degrees of the LAG and PSG Hierarchies

	LA Grammar	PS Grammar
undecidable	—	recursively enumerable languages
decidable	A languages[4]	—
exponential	B languages	context-sensitive languages
exponential	C3 languages	
polynomial	C2 languages	context-free languages
linear	C1 languages	regular languages

The LAG hierarchy does not have a class of undecidable languages, while the PSG hierarchy does not have a class of decidable languages. The statement 'there is a parser for the context-free languages' means that there is a PSG parser which can handle all languages in the class, e.g., the Earley parser. 'There is no parser for the context-sensitive languages' means that there is no parser for all languages in the class. Thus, there may exist a computationally tractable parser specifically for context-sensitive $a^n b^n c^n$, but not for inherently complex 3SAT or SUBSET-SUM, which are also context-sensitive.

13.4 Sub-Hierarchy of C1, C2, and C3 Lags

Compared to the A and B LAGs, the C LAGs constitute the most restricted class of LAGs, parsing the smallest LAG class of languages. However, compared to the context-free languages (which are properly contained in the C languages), the class of C languages is quite large (13.2.1). It is therefore theoretically interesting and practically useful to differentiate the C LAGs further into subclasses by defining a sub-hierarchy.

In the C LAGs, the complexity of a rule application is constant (TCS'92, Definition 4.1). Therefore, the number of rule applications in a C LAG derivation depends solely on the ambiguity degree. Different ambiguity degrees naturally define the sub-hierarchy of the C LAGs: in the subclasses of C1, C2, and C3 LAGs, increasing degrees of ambiguity result in increasing degrees of complexity.

The subclass with the lowest complexity and the lowest generative capacity is the C1 LAGs. A C LAG is a C1 LAG if it is not recursively ambiguous. The class of C1 languages parses in linear time and contains all deterministic context-free languages which are recognized by a DPDA without ε-moves, plus context-free languages with –recursive ambiguities, e.g., $a^k b^k c^m d^m \cup a^k b^m c^m d^k$, as well as many context-sensitive languages, e.g., $a^k b^k c^k$, $a^k b^k c^k d^k e^k$, $\{a^k b^k c^k\}^*$, L_{square}, L_{hast}^k, a^{2^i}, $a^k b^m c^{k \cdot m}$, and $a^{i!}$, whereby the last one is not even an index language.[5] Examples of

---

[4] The algebraic definition of LA-grammar (FoCL 10.2) benefited greatly from help by Professor Dana Scott, who also provided the proof that the class of A-languages comprises *all* recursive languages (FoCL 11.1.3).

unambiguous context-sensitive C1 LAGs are $a^k b^k c^k$ defined in 10.3.3 and $a^{2^i} =_{def} \{a^i \mid i$ is a positive power of 2$\}$ (TCS'92 Definition 5.) A non-recursively ambiguous C1 LAG is $a^k b^k c^m d^m \cup a^k b^m c^m d^k$ (TCS'92)[6].

A C LAG is a C2 LAG if it generates recursive ambiguities which are restricted by the single return principle.

13.4.1 THE SINGLE RETURN PRINCIPLE (SRP)

A recursive ambiguity is single return if exactly one of the parallel paths returns to the state resulting in the ambiguity in question.

The class of C2 languages parses in polynomial time and contains certain nondeterministic context-free languages like WW^R and L^∞_{hast}, plus context-sensitive languages like WW, $W^{k \geq 3}$, $\{WWW\}^*$, and $W_1 W_2 W_1^R W_2^R$.[7]

For example, the worst case in parsing WW^R is inputs consisting of an even number of the same word (letter). Consider the derivation structure for the input a a a a a a, with 1 for r-1 and 2 for r-2.

13.4.2 DERIVATION STRUCTURE OF THE WORST CASE IN WW^R

rules:	applications:
2	a\$a
122	aa\$aa
11222	aaa\$aaa
11122	aaaa\$aa
11112	aaaaa\$a
11111	aaaaaa\$

The unmarked middle of the intermediate strings generated in the course of the derivation is indicated by \$. Of the six hypotheses, the first two are invalidated by the fact that the input string continues, the third hypothesis correctly corresponds to the input a a a a a a, and the remaining three hypotheses are invalidated by the fact that the input does not provide any more words (grammatical disambiguation in a formal language).

A C LAG is a C3 LAG if it generates unrestricted recursive ambiguities. The class of C3 languages parses in exponential time and contains the deterministic context-free language L_{no}, the hardest context-free language HCFL, plus context-sensitive languages like SubsetSum and SAT, which are $\mathcal{NP}$-complete.[8]

---

[5] A C1 LAG for $a^k b^k c^m d^m \cup a^k b^m c^m d^k$ is defined in FoCL 11.5.2; for L_{square} and L^k_{hast} in Stubert (1993), pp. 16 and 12; for $a^k b^k c^k d^k e^k$ in CoL, p. 233; for $a^k b^m c^{k \cdot m}$ in TCS'92, p. 296 and for a^{2^i} in FoCL 11.5.1. A C1 LAG for $a^{i!}$ is sketched in TCS'92, p. 296, footnote 13.

[6] This language has been called *inherently ambiguous* because there is no unambiguous PSG for it (Hopcroft and Ullman 1979, pp. 99–103).

[7] The C2 LAGs for WW^R and WW are defined in 13.3.3; for L^∞_{hast} in Stubert 1993, p. 16; for WWW in CoL, p. 215; for $W^{k \geq 3}$ in CoL, p. 216; and for $W_1 W_2 W_1^R W_2^R$ in FoCL 11.5.7.

[8] A C3 LAG for L_{no} is defined in 13.3.5; for HCFL in Stubert (1993), p. 16; for SubsetSum in FoCL 11.5.8; and for SAT in TCS'92, p. 302, footnote 19.

13.5 Applying LAG to Natural Language

In a LAG, the lexical entries have a two-level structure, consisting of a surface and a category, e.g., [a (a)] in linear notation. During a derivation, the surface and the category may diverge, e.g., [aaab (aab)] (FoCL 10.4.1, segment 4). In a LAG for a formal language, the rules have abstract names like r-1 or r-2.

In the application of LAG to natural language, the possible divergence between surfaces and categories is used for defining grammatically motivated categories such as [gave (N' D' A' V)]. Also, the abstract rule names are replaced by grammatically meaningful ones like DET+CN or NP+VERB.

The following analysis is based on the LAG *LA E2* defined in FoCL 17.4.1 for a small fragment of English:

13.5.1 TIME-LINEAR LAG ANALYSIS OF AN ENGLISH SENTENCE

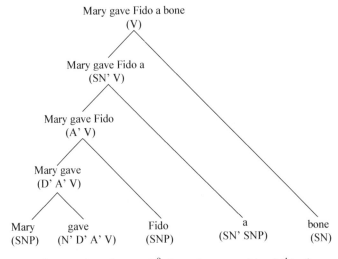

The first word [Mary (SNP)] cancels[9] the valency position N' in the second word [gave (N' D' A' V)], resulting in the new sentence start [Mary gave (D' A' V)] one level above (bottom up). Next the current sentence start [Mary gave (D' A' V)] combines with the current next word [Fido (SNP)], resulting in the new sentence start [Mary gave Fido (A' V)] with the canceled valency position D'. This time-linear procedure continues until all valency positions are canceled, resulting in [Mary gave Fido a bone (V)].

The application of this method to 221 constructions of German and 114 constructions of English in NEWCAT'86 showed that a strictly time-linear derivation order for natural language was empirically feasible. Also, the NEWCAT program was shown to be extremely efficient computationally as compared to competing efforts at the time, such as phrase structure-based LFG.

Nevertheless, NEWCAT is still a stand-alone algorithm in the style of classic complexity analysis. Instead of being data-driven, the system of rule packages in a LAG constitutes a finite state transition network (FSN). By annotating the transitions with the associated rule name, the FSN is turned into an ATN.

Consider the ATN of the LAG grammar for context-sensitive $a^n b^n c^n$ (13.3.2):

13.5.2 ANNOTATED TRANSITION NETWORK OF THE LAG FOR $a^n b^n c^n$

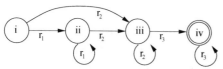

This ATN consists of four states, represented as the circles i – iv. Each state is defined as an ordered pair consisting of a rule name and a rule package. State i corresponds to the start state ST_S, while the states ii, iii, and iv correspond to the output of rules r_1, r_2, and r_3. State iv has a double circle, indicating a possible final state (definition of ST_F in 13.3.2).

However, when the LAG system was applied to basic structures of natural language, such as extending declaratives to yes-no interrogatives, it turned out that the use of rule packages became prohibitively complex. Consider the finite state backbone of *LA E3* defined in FoCL 17.5.6.

13.5.3 ANNOTATED TRANSITION NETWORK FOR *LA E3*

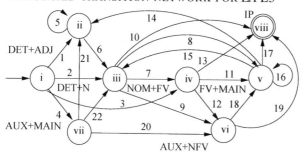

(ii)	1, 5, 14, 21	DET+ADJ		(v)	8, 11, 16, 18	FVERB+MAIN
(iii)	2, 6, 15, 22	DET+N		(vi)	9, 12, 20	AUX+NFV
(iv)	3, 7	NOM+FV		(vii)	4	AUX+MAIN
				(viii)	10, 13, 17, 19	IP

Including the start state, there are eight states. As in 13.5.2, the transition from one state to the next is annotated with a rule name, thus restricting the transition to a specific categorial operation. The problem was not in writing the rule packages for controlling the parsing of a particular natural language construction, but writing the ATN for the complete system.

---

[9] At this stage of the theory, valency positions are canceled by deletion, as in CG. This is not a problem as long as the theory is sign-based. In DBS, however, the valency information must be preserved for repeated hear say in the speak mode by using #-marking for canceling. For example, [Mary gave (D′ A′ V)] cancels by deletion in LAG, but in DBS by #-marking, as in [Mary gave (#N′ D′ A′ V)].

It turned out that such ATNs do not provide the hoped for contribution to heuristics. This coincided with a more general problem inherited from classic complexity theory, namely the stand-alone nature of LAG as an algorithm.

13.6 From LAG to the Hear Mode

LAG and DBS share the time-linear derivation order, but differ in their method of loading the input, which is *holistic* in LAG and *incremental* in DBS. Holistic loading takes a complete sequence as input, e.g., aaabbbccc, and then processes it word by word from left to right. This works for parsing individual sentences in a collection of linguistic examples, but is impractical for texts like a Tolstoy novel.

Processing in a holistic loading system may be illustrated as follows:

13.6.1 CONNECTING A SENTENCE START TO ITS SUCCESSOR

surface level aa   abbbbccc$\Rightarrow$ aaa bbbccc
rule level r_1:(X) (a) $\Rightarrow$ (aX) $\{r_1, r_2\}$

At the surface level, the next word is added at the end of the current sentence start, regardless of the categorial operation. At the rule level, the category of the sentence start is (X) and the category of the next word is (a). In 13.3.2, rule r_1 attaches the next word category a at the front end of the sentence start variable (X) as (aX) and calls the rule package $\{r_1, r_2\}$. The rules of the rule package are applied to the resulting sentence start and the new next word in the loaded input sequence.

In the hear mode, the start of parsing is a special case because an initial operation activated by a proplet matching its second input pattern can not find a proplet matching its first input proplet at an empty now front. In a text, the initial composition is unique, but in the linguistic analysis of isolated examples each has one.

For example, the derivation of The dog barked. as an isolated linguistic example begins with the recognition of the surface The. Based on matching a type on raw data provided by the input component, it serves as input to automatic word form recognition. The output is stored at the now front, which happens to be empty:

13.6.2 STORING SENTENCE-INITIAL WORD AT EMPTY NOW FRONT

member proplets   *now front*   owners
$\begin{bmatrix} \text{sur: Der} \\ \text{noun: N\_}n \\ \text{cat: CN}' \text{ NP} \\ \text{sem: sg} \\ \text{fnc:} \\ \ldots \\ \text{prn: 23} \end{bmatrix}$ the

Without a next word yet, no operation is activated and the derivation continues with another automatic word form recognition, resulting in the following constellation:

13.6.3 STORING NEXT WORD AT NOW FRONT

member proplets	*now front*	*owners*
	$\begin{bmatrix} \text{sur: Hund} \\ \text{noun: dog} \\ \text{cat: sn} \\ \text{sem: sg} \\ \text{fnc:} \\ \text{mdr:} \\ \text{nc:} \\ \text{pc:} \\ \text{prn: 23} \end{bmatrix}$	dog

...

| | $\begin{bmatrix} \text{sur: Der} \\ \text{noun: N\_}n \\ \text{cat: CN}'\text{ NP} \\ \text{sem: sg} \\ \text{fnc:} \\ \text{mdr:} \\ \text{nc:} \\ \text{pc:} \\ \text{prn: 23} \end{bmatrix}$ | the |

The token line for storing a next word at the now front is determined alphabetically by the core value, here dog. This supports efficient computational string search (Knuth et al. 1977) for storage and retrieval.

From here on out, the derivation continues in standard fashion. The 'next word,' here a noun, activates all recognition operations which match it with their second input pattern (operations 28–53 in TExer). Activated operations look at the now front for input matching their first input pattern. Those which find one apply.

13.6.4 ABSORBING *dog* INTO *the* WITH DET∪CN

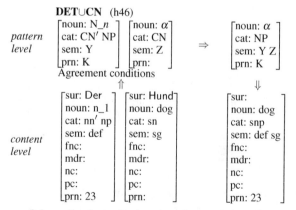

The successful application of a hear mode operation triggers the lookup and storage of another 'next word,' here bark, by automatic word form recognition:

13.6.5 STORING NEXT WORD AT CURRENT NOW FRONT

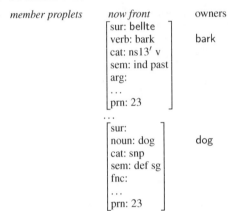

The storage of a next word automatically activates all operations which match the next word with their second input pattern (operations 1–27 in TExer).

13.6.6 CROSS-COPYING *dog* AND bark WITH SBJ×PRD

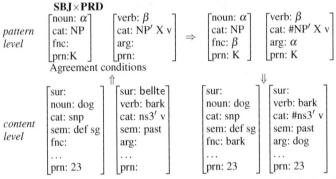

Depending on the input, the derivation may continue indefinitely, for example by adding interpunctuation and going on to the next sentence, resulting in a content defined as a set (order-free) of proplets connected and ordered by address, e.g., (bark 23). With the start of a new sentence, the accumulation of parallel readings, if any, starts from scratch.

13.7 From the Hear Mode to the Speak Mode

The counterpart of the DBS hear mode is the speak mode. It rides piggyback on the think mode which activates content by navigating along the semantic relations of structure coded by address. A speak mode derivation is a think mode navigation with the optional production of language-dependent surfaces.

A think-speak mode operation has one input and one output pattern. The operators are $\swarrow$ and $\nearrow$ for the subject/predicate, $\searrow$ and $\nwarrow$ for the object\predicate, $\downarrow$ and $\uparrow$ for the modifier|modified, and $\rightarrow$ and $\leftarrow$ for the conjunct−conjunct relation. Language-dependent surfaces are produced from the goal proplet.

The following speak mode production uses the content derived in 13.6 as input:

13.7.1 NAVIGATING WITH V/N FROM *bark* TO *dog*

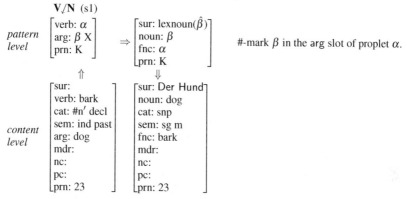

The language-dependent surface is realized from a list which connects relatively language-independent core values to their language-dependent counterpart, here dog $\Rightarrow$ Hund, and interprets the cat value def and the sem value sg as the German definite article form der. The resulting surface is Der Hund.

The next navigation step returns from the subject to the verb:

13.7.2 NAVIGATING WITH N/V FROM dog BACK TO *bark*

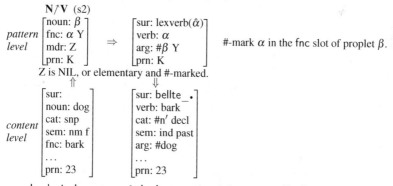

Lexverb uses bark, ind past, and decl to produce the surface bell-te_.. In LAG, the rules to be tried are called by the rule package of the current rule. In DBS, in contrast, the operations to be tried are activated by the next word (data-driven).

13.8 Incremental Lexical Lookup in the Hear Mode

The next word originates as raw data input to a sensor of the agent's interface component and is recognized as a language-dependent surface. The surface is used for lexical lookup, the result of which is stored at the current now front (CC 12.4.4):

13.8.1 OWNER-BASED STORAGE OF LANGUAGE PROPLET AT NOW FRONT

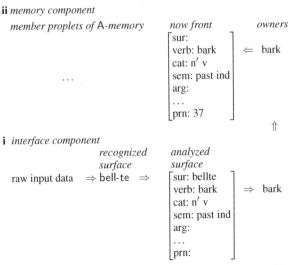

At level (i), a language-dependent word form type matching the raw data results in the recognized surface (⇒) bell-te (here letter sequence). It is used for lexical lookup of the analyzed surface, i.e., the complete proplet, from the allomorph trie structure (CC 12.5.3). Using string search, the core value bark serves (a) to access (⇑) the token line of bark and (b) to store (⇐) the proplet retrieved from the trie structure, without the sur value but with an automatically assigned prn value, at the now front (ii).

Each activated operation looks at the now front for a proplet matching its first input pattern. In natural language parsing, the number of proplets at the current now front is usually no more than four or five because the now front is cleared whenever a proposition (subclause) is completed, indicated by in- or decrementing the prn value (CLaTR² 11.4.10). After processing, now front clearance leaves the proplets behind as member proplets (loomlike clearance). The now front is cleared by moving it and the owner values one step to the right into fresh memory space.

Content stored as member proplets cannot be changed. The only way to correct is adding new content, like a diary entry referring by address to the content to be corrected. Proplets without an open continuation slot are not tried as input to a first input pattern. The only way to increase the DBS hear mode complexity above linear is a *recursive* ambiguity, which is prevented in natural language by grammatical disambiguation.

13.9 Ambiguity in Natural Language

There is repeating ambiguity in natural language, but to increase complexity, the readings would have to be [+global] (FoCL 11.3). In natural language, systematically repeating[10] [+global] ambiguity does not seem to exist. Consider, for example, the following derivation structure:[11]

13.9.1 AMBIGUITY STRUCTURE OF AN UNBOUNDED SUSPENSION

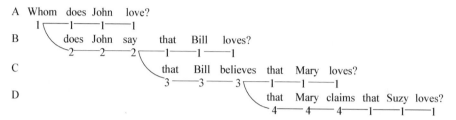

In line A, who(m) is the object of an elementary proposition with a transitive verb which does not take a clausal object. Thus all proplets in line A share the prn value 1. In line B, in contrast, the matrix verb takes a clausal object which who(m) belongs to. Thus, does John say X has the prn value 2, whereby X is Bill loves who(m) with the prn value 1. The construction in line B terminates because the verb *love* does not take a clausal object.

Line C branches off line B because the first object clause uses a verb which takes a second object clause as its oblique argument. Thus, does John say X continues to have the prn value 2, but the new object clause Bill believes Y has the new prn value 3, whereby Y is that Mary loves who(m) with the prn value 1. The construction terminates in branch C.

Line D branches off C because the second object clause uses a verb which takes a third object clause as its argument. Thus, does John say X continues to have the prn value 2, Bill believes Y continues to have the prn value 3, but the new object clause that Mary claims Z has the prn value 4, whereby Z is that Suzy loves who(m) with the prn value 1.

Even though an unbounded suspension (i) may be continued indefinitely and (ii) causes a systematic syntactic ambiguity, it does not increase the computational complexity of natural language (FoCL 11.5). This is because one of the two branches always terminates before the next ambiguity is complete. In other words, there is no global ambiguity in 13.9.1, in the same sense as there is no global ambiguity in the 'garden path' sentence (FoCL 11.3.6).

Another construction with a repeating local ambiguity is adnominal (aka relative) clauses. See TExer3 Sects. 3.3, 3.4, and 5.6 for complete declarative analyses. As long as no natural language can be shown to have repeating global ambiguity, the Linear Complexity Hypothesis for natural language remains without counterexample.

---

[10] Systematically repeating (i.e. recursive or iterative) [+global] readings would be a serious impediment to successful communication.

[11] For the complete declarative DBS analysis of this example see TExer 5.5.

13.10   Language Dependence of Grammatical Disambiguation

Ambiguities are language dependent. For example, the translation of *flying airplanes* into German disambiguates the English readings grammatically into fliegende Flugzeuge and Flugzeuge fliegen. The translation of *horse raced by the barn* into German disambiguates the English readings into Pferd jagte ... vorbei and Pferd das ... vorbeigejagt wurde. The translation of *man who* into German disambiguates the English readings into Mann der and Mann den. English *Who does John say that ...* doesn't even have a literal translation into German. With English as an isolating and German as an inflectional language, local ambiguities and their grammatical resolution seem to be a typological phenomenon.

13.11   Bach-Peters Sentence

The computational undecidability of natural language[12] alleged by Phrase Structure Grammar (PSG) is based on a transformational analysis of the following example:

13.11.1 THE BACH-PETERS SENTENCE

THE MAN WHO DESERVES *it* WILL GET THE PRIZE *he* WANTS

The formal proof by Peters and Ritchie (1973) relies on two reciprocal recursions, one deriving the pronoun it transformationally from the 'full' noun phrase the prize he wants, the other deriving the pronoun he transformationally from the 'full' noun phrase man who deserves it.

The alternative DBS hear mode analysis (CLaTR^2 11.4.9–11.4.12) is of linear time complexity because the coreference between the prize he wants and it is defined by address instead of a transformation, and similarly for the coreference between man who deserves it and he. As in natural language communication, ambiguity in DBS is limited to the hear mode. The speak mode counterpart to hear mode ambiguity is paraphrase. While the hearer's ambiguity may result in multiple simultaneous readings, paraphrase is a matter of choice which depends on the speaker's rhetorical purpose. Paraphrases are of linear complexity.

Hear mode ambiguity is of two kinds, [+global] and [-global] (FoCL 11.3). Relevant for complexity are only the [+global] ambiguities. Because each reading of length n requires exactly n-1 derivation steps and each derivation step of the DBS C-LAGs are below a grammar-dependent constant C, the computational complexity of the hear mode depends solely on the number of readings.

---

[12] Classifying natural language as computationally undecidable has been noted to be unlikely by Harman (1963), Gazdar (1981), McCawley (1982), Ross (1986), and many others.

13.12   Conclusion

This chapter compares the computational complexity of three algorithms for analyzing natural language: (i) the sign-based substitution-driven algorithm of PSG, (ii) the sign-based data-driven algorithm of LAG, and (iii) the agent-based data-driven speak and hear mode algorithms of DBS.

Sign-based PSG and LAG have in common that they do not distinguish between the speak and the hear mode, but differ in their derivation principle, which is substitution-driven in PSG, but data-driven in LAG. More specifically, the input to a derivation in PSG, i.e. ST, EST, REST, GB, GPSG, HPSG, etc., is always the same S node (for start or sentence) and the output is different phrase structures. The input to a LAG derivation, in contrast, is a time-linear sequence of lexical proplets provided by automatic word form recognition, and the output a content in which the lexical proplets are connected by the classical semantic relations of structure, i.e. subject/predicate, object\predicate, modifier|modified, and conjunct–conjunct.

Data-driven LAG and DBS have in common that they compute possible continuations, but differ in that DBS distinguishes between the speak and the hear mode, while LAG does not. In agent-based DBS, the input to the speak mode is a content, defined as a set of proplets connected by the classical semantic relations of structure, and the output a language-dependent surface. The input to the hear mode is a language-dependent surface and the output a content.

Summary:
Ambiguity in natural language communication is limited to hear mode interpretation and may result in multiple simultaneous readings. The speak mode counterpart to ambiguity is paraphrase. For speak mode production, alternative paraphrases require a choice, guided by speaker's rhetorical purpose. Each paraphrase is by principle unambiguous and therefore of linear complexity.

Hear mode ambiguity is of two kinds, [+global] and [-global] (FoCL 11.3). Relevant for computational complexity are the [+global] ambiguities. Because each reading of length n requires exactly n-1 derivation steps and each derivation step of the C-LAGs is by definition below a grammar-dependent constant C, the computational complexity of the DBS hear mode depends solely on the number of readings.

Because recursive ambiguity is prevented by grammatical disambiguation (13.9.1), the hear mode is of linear complexity, like the speak mode. Therefore the overall computational complexity degree of natural language is linear.

14. Database Semantics vs. Predicate Calculus

A content like Fido found a bone. may be viewed from two basic perspectives. One takes the view of an outside observer. For example, Predicate Calculus analyzes the truth relation between a formal sign and a formal model (14.1) or model structure (14.8). This approach is called *sign-based*.

The other takes the view of the agent, i.e. Fido looking, listening, and sniffing out into the world, keeping track of it, and deriving suitable action. For example, DBS models the cognition of an artificial agent with an on-board interface component for automatically monitoring recognition (14.11) and action (14.12), and a content-addressable, on-board memory component for the storage and retrieval of content. Because the agent's processing of sensory input and output is central to this approach, it is called *agent-based*.[1]

14.1 Definition of Predicate Calculus

Montague's (1973) version of first order Predicate Calculus, hence abbreviated PredC, is widely admired.[2] Leaving the intension-extension distinction and the use of lambda calculus aside, it may be presented as follows (slightly revised):

14.1.1 FORMAL DEFINITION OF PREDC

1. A model $\mathscr{M}$ is defined as the quadruple $<A, B, F, g>$, where A is an infinite set of objects or individuals, B a finite set of basic expressions, F a denotation function from B into the free monoid[3] A^* over A, and g an assignment function from variables into A^*.
2. The elements $\emptyset$ (empty set) and $\{\emptyset\}$ (set containing the empty set) of A^* are used as the denotation of the truth values 1 (true) and 0 (false), respectively.
3. Syntactically, the operators $\neg, \wedge, \vee, =, \rightarrow, \forall$, and $\exists$ are defined as follows:
   a) If f is a one-place functor and α is a name, then $f(\alpha)$ is a sentence.
   b) If ϕ is a sentence, then $\neg\phi$ is a sentence.

---

[1] For earlier attempts at integrating symbolic logic into an agent-based cognition see Hausser (1980), Kamp (1980), and Barwise and Perry (1983).

[2] Thanks to Professors Nuel Belnap, Georg Kreisel, and Rich Thomason for deepening my understanding of formal semantics and Montague Grammar.

[3] The * is called the Kleene Star. The free monoid A^* serves as the "universe of discourse."

© The Author(s), under exclusive license to Springer Nature Switzerland AG 2023
R. Hausser, *Ontology of Communication*, https://doi.org/10.1007/978-3-031-22739-4_14

    c) If ϕ is a sentence and ψ is a sentence, then $\phi \wedge \psi$ is a sentence.
    d) If ϕ is a sentence and ψ is a sentence, then $\phi \vee \psi$ is a sentence.
    e) If ϕ is a sentence and ψ is a sentence, the $\phi \rightarrow \psi$ is a sentence.
    f) If ϕ is a sentence and ψ is a sentence, then $\phi = \psi$ is a sentence.
    g) If f and h are functors and x is a variable, then $\exists x[f(x) \wedge h(x)]$ is a sentence.
    h) If f and h are functors and x is a variable, then $\forall x[f(x) \rightarrow h(x)]$ is a sentence.

Definitions (a–f) constitute Propositional Calculus, hence abbreviated PropC. Definitions (g–h) extend PropC into PredC by introducing the quantifiers $\exists x$ and $\forall x$ binding a variable. Within a formula, the *scope* of a quantifier is the area in which its variable is bound; the area is defined by the formula's bracketing structure.

4. Semantically, the set of operators is defined as follows:
    a) If f is a one-place functor, then $f(\alpha)$ is a true sentence relative to a model $\mathcal{M}$
       iff if the denotation of α in $\mathcal{M}$ is an element of the denotation of f in $\mathcal{M}$.
    b) $\neg\ \phi$ is a true sentence relative to a model $\mathcal{M}$ if and only if the denotation of
       ϕ is 0 relative to $\mathcal{M}$.
    c) $\phi \wedge \psi$ is a true sentence relative to a model $\mathcal{M}$ if and only if the denotations
       of ϕ and of ψ are 1 relative to $\mathcal{M}$.
    d) $\phi \vee \psi$ is a true sentence relative to a model $\mathcal{M}$ if and only if the denotation of
       ϕ or ψ is 1 relative to $\mathcal{M}$.
    e) $\phi \rightarrow \psi$ is a true sentence relative to a model $\mathcal{M}$ iff the denotation of ϕ is 0
       relative to $\mathcal{M}$ or the denotation of ψ is 1 relative to $\mathcal{M}$.
    f) $\phi = \psi$ is a true sentence relative to a model $\mathcal{M}$ if and only if the denotation
       of ϕ relative to $\mathcal{M}$ equals the denotation of ψ relative to $\mathcal{M}$.
    g) $\exists x[f(x) \wedge h(x)]$ is a true sentence relative to $\mathcal{M}$ and a variable assignment g iff
       at least one $g'(x)$ makes $[f(x) \wedge h(x)]^{\mathcal{M},g}$ true.
    h) $\forall x[f(x) \rightarrow h(x)]$ is a true sentence relative to $\mathcal{M}$ and a variable assignment g if
       and only if all $g(x)$ make $[f(x) \rightarrow h(x)]^{\mathcal{M},g}$ true.

Today's PredC originated with Frege (1879). It was complemented by Montague with the assignment g, lambda calculus, and the intension-extension dichotomy. In DBS, the assignment g is replaced by substitution variables with simultaneous substitution, the role of lambda reduction is taken by time-linear derivations in the speak and the hear mode, and instead of the intension-extension dichotomy (based on possible worlds) DBS uses the type-token relation, implemented as efficient computational pattern matching based on nonrecursive feature structures with ordered attributes, called proplets.

14.2 PredC Overgeneration

The syntactic rules of PropC and their semantic interpretation rely on recursion. For example, if $(p \wedge q)$ and r are sentences, then $((p \wedge q) \wedge r)$, $((p \wedge q) \vee r)$, $((p \wedge q) \rightarrow r)$, and $((p \wedge q) = r)$ as well as $(r \wedge (p \wedge q))$, $(r \vee (p \wedge q))$, $(r \rightarrow (p \wedge$

q)), and $(r = (p \wedge q))$ are also sentences. The (i) order of the sub-sentences does not make a difference semantically (symmetry) except for the operator '→' (asymmetry), the (ii) sentences resulting from the variations all have well-defined semantic interpretations relative to $\mathcal{M}$, and (iii) there is no upper bound on the number of variations because there is no limit on the length of PredC formulas.

It is similar for $\forall x[f(x) \rightarrow h(x)]$ and $\exists y[i(y) \wedge j(y)]$ of PredC, except that (i) the semantic interpretation relative to a model $\mathcal{M}$ requires the variable assignment g, (ii) functors may take more than one quantified variable as an argument, e.g., $f(x,y)$ or $f(x,y,z)$, and (iii) alternative orders of substitution may produce a systematic ambiguity such as the following:

14.2.1 ALLEGED AMBIGUITY OF Every dog finds a bone

1. $\forall x[dog(x) \rightarrow \exists y[bone(y) \wedge find(x,y)]]$
   Every dog finds a bone
2. $\exists y[bone(y) \wedge \forall x[dog(x) \rightarrow find(x, y)]]$
   There exists a bone which every dog finds

When PredC is used as the ready-made semantic interpretation of an independently motivated ("autonomous") syntax, specifically "innate" PSG, both readings are automatically assigned. Thereby reading (1) is an intuitively correct representation of the meaning but reading (2) is not. Without being a genuine paraphrase, the PrepC structure of reading (2) is an artifact of the formalism. Because the alleged ambiguity occurs whenever the $\exists$ and $\forall$ quantifiers appear in the same formula, it is a systematic overgeneration of PredC.

14.3 Determiners

The DBS alternative to the PredC formulas in 14.2.1 is a single set of proplets, connected by semantic relations of structure, coded by address.

14.3.1 UNAMBIGUOUS DBS ANALYSIS OF Every dog finds a bone

sur:	sur:	sur:
noun: dog	verb: find	noun: bone
cat: snp	cat: #ns3 #a decl	cat: snp
sem: pl exh	sem: pres	sem: indef sg
fnc: find	arg: dog bone	fnc: find
mdr:	mdr:	mdr:
nc:	nc:	nc:
pc:	pc:	pc:
prn: 4	prn: 4	prn: 4

As in natural language, but unlike 14.2.1, there is no ambiguity, no coordination, and no coreference. The determiner aspect of $\forall x$ representing every in 14.2.1 is coded as the features [cat: snp] (singular noun phrase) and [sem: pl exh] (plural exhaustive) of the *dog* proplet, while that of ∃y representing a(n) is coded as the features [cat: snp] (singular noun phrase) and [sem: sg indef] (singular indefinite) of the *bone* proplet. The semantic relations between the proplets *dog* and *find* are subject/predicate, and between *find* and *bone* object\predicate.

14.4 PredC Undergeneration

From a linguistic point of view, the counterpart to PredC overgeneration is PredC undergeneration, i.e. certain meaningful grammatical constructions cannot be properly expressed. The classic example is the *donkey sentence*[4]: Every farmer who owns a donkey beats it (Geach 1962). The PredC derivation is driven by systematic substitution and results in the following formula:

14.4.1 INCORRECT ANALYSIS OF WELL-FORMED SENTENCE IN PREDC

> Every farmer who has a donkey beats it
> $\forall x[[\text{farmer}(x) \wedge \exists y[\text{donkey}(y) \wedge \text{own}(x,y)]] \rightarrow \text{beat}(x,y)]$

The English sentence is grammatical and meaningful, and the substitutions of the derivation are correct, but the resulting formula is semantically inappropriate because the variable y in beat(x,y) is not in the scope of the quantifier ∃y binding donkey(y).[5]

14.5 Coreference by Address

Alternative to treating coreference by means of quantifiers binding variables, DBS treats all semantic relations of structure, including coreference (CC 6.5), by means of proplet-internal address values:

14.5.1 INTERPRETATION OF THE DONKEY SENTENCE IN DBS

⌈noun: farmer⌉	⌈verb: own⌉	⌈noun: **donkey**⌉	⌈verb: beat⌉	⌈noun: **(donkey 17)**⌉
cat: snp	cat: #n′ #a′ v	cat: snp	cat: #ns3′ #a′ decl	cat: snp
sem: pl exh	sem: pres	sem: indef sg	sem: pres	sem: sg
fnc: beat	arg: ∅ donkey	fnc: own	arg: farmer (donkey 17)	fnc: beat
mdr: (own 17)	mdd: (farmer 16)	mdr:	mdr:	mdr:
nc:	nc:	nc:	nc:	nc:
pc:	pc:	pc:	pc:	pc:
⌊prn: 16⌋	⌊prn: 17⌋	⌊prn: 17⌋	⌊prn: 16⌋	⌊prn: 16⌋

---

[4] The quantifier scope problem manifested by the donkey sentence was recognized in the middle ages (Walter Burley 1328/Gualterus Burlaeus 1988). It is one of several instances in which PredC can not provide the semantically correct quantifier scope. Discourse Representation Theory (DRT, Kamp 1980, Kamp and Reyle 1993) attempted to solve the problem of the donkey sentence while trying to maintain the sign-based substitution-driven foundation of PredC.

The grammatical embedding is reflected by the prn values, which are automatically incremented from 16 to 17 in the transition from the main to the sub-clause and decremented from 17 to 16 in the transition back to the main clause. The coreferential pronoun it in the main clause is represented by the extrapropositional address (donkey 17), which refers to the antecedent[6] in the sub-clause.

14.6 In PredC, Propositions Denote Truth Values

Another important difference between DBS and PredC is the respective treatment of concepts like dog, small, find, bone, and big. In PredC, concepts are treated as elementary mini-propositions which denote truth values relative to a set-theoretic model. Formally defined as functors which may differ in the number of arguments, e.g., $f(x)$ vs. $f(x,y)$, they are connected by the propositional operators of PropC and the quantifiers of PredC:

14.6.1 NOUN, VERB, AND ADJ FLATTENED INTO MINI-PROPOSITIONS

the noun dog is interpreted as x is a dog and written as $dog'(x)$
the adj little is interpreted as x is little and written as $little'(x)$
the 1-place verb snore is interpreted as x snores and written as $snore'(x)$ the
2-place verb find is interpreted as x finds y and written as $find'(x, y)$
the 3-pl. verb give is interpreted as x gives y z and written as $give'(x, y, z)$

This allows to represent, for example, The little dog found a big bone as five mini-propositions which are coordinated with the propositional operator $\land$ and have the variables x and y bound by two $\exists$ quantifiers:

14.6.2 PREDC REPRESENTATION OF The little dog found a big bone

$$\exists x[dog'(x) \land little'(x) \land \exists y[bone'(y) \land big'(y) \land find'(x, y)]]$$

The meaning difference between the constants dog, little, bone, big, and find depends on the denotation function F and the assignment function g, provided they are explicitly defined by the logician, which is usually not the case. The reason may be shown by explicitly defining a possible model for a PredC formula:

14.6.3 MINIMAL MODEL FOR THE PREDC FORMULA 14.6.2

Let $\mathcal{M}$ be a model <A, B, F, g>, where A is an infinite set of objects or individuals, B a finite set of basic expressions, F a denotation function from

---

[5] There have been numerous proposals to avoid the "dangling variable" by fronting the existential quantifier. In this way, the y in donkey(y) would get bound by $\exists y$ but at the cost of losing compositionality, which is methodologically unacceptable.

[6] In linguistics, the term 'antecedent' is used not only for inferences, but also for the full noun referent, here a donkey, preceding a coreferential pronoun, here it (CLaTR 11).

B into A*, and g an assignment function from variables into A*.

For illustration, let us define A, B, F, and g as follows:
$A = \{a_1, a_2, a_3\}$
$B = \{dog, small, big, bone, eat\}$
$F(dog) = a_1, F(small) = \{a_1\}, F(big) = \{a_2\}, F(bone) = a_2, F(eat) = <a_1 \, a_2>,$
$g(x) = a_1, g(y) = a_2$

Based on definitions in 14.6.3, the formula 14.6.2 is well-formed and true in $\mathscr{M}$. However, if we defined $F(dog)$ as a_3, for example, the formula would be false.

In summary, because PredC's precomputational ontology provides neither a memory nor an interface component for autonomous recognition and action, formal models like 14.6.3 must be defined by hand, though a complete modeling of even the tiniest part of the world (i) is out of reach and (ii) without any practical purpose. Therefore, explicit models are almost never defined. Instead logicians circumscribe models by means of conditionals which encode their intuitions about truth. DBS, in contrast, treats the world surrounding the agent's cognition *as given*[7] and limits itself to automatically monitor the agent's recognition and action computationally, using the agent's on-board interface and memory components.

14.7 In Database Semantics, Propositions Are Content

From a linguistic point of view, treating nouns, adjs, and verbs uniformly as minipropositions (14.6.1) is semantically misguided because (i) it loses the classical distinction between referents, properties, and relations (CC 1.5.3), and (ii) obscures the empirical fact that properties and relations do not refer. Also, the use of coordination and coreference in connection with the quantifiers $\forall$ and $\exists$ (iii) violates the methodological standard of surface compositionality (FoCL 4.5) because there is neither coordination nor coreference in natural language expressions such as The little dog found a big bone.

DBS, in contrast, differentiates concepts into the three semantic kinds *referent, property*, and *relation*, with the syntactic correlates of elementary *noun, adj*, and *verb* (CC 1.5). Also, instead of propositions "denoting truth-values," as in PropC and PredC, propositions 'are content' in DBS:

14.7.1 THE CONTENT OF The little dog found a big bone

sur:	sur:	sur:	sur:	sur:
noun: **dog**	adj: **little**	verb: **find**	adj: **big**	noun: **bone**
cat: def sg	cat: adn	cat: #n′ #a′ decl	cat: adn	cat: def sg
sem: animal	sem: pad	sem: ind past	sem: pad	sem: dog food
fnc: **find**	mdd: **dog**	arg: **dog bone**	mdd: **bone**	fnc: **find**
mdr: **little**	mdr:	mdr:	mdr:	mdr: **big**
nc:	nc:	nc:	nc:	nc:
pc:	pc:	pc:	pc:	pc:
prn: 47	prn: 47	prn: 47	prn: 47	prn: 47

The proplets of a content are a set (order-free), held together by (i) a common prn value, here 47, and (ii) the semantic relations of structure coded by address (CLaTR 4.4, 11).

In addition to coding a semantic relation from one proplet to another, an address specifies a proplet's unique *storage location* (CC 12.4.4) in the agent's content-addressable A-memory (CC 2.3). The core value of the address is used for finding the token line (vertical) and the prn value for finding the position in the token line (horizontal). Computationally, the token line is found by the letter sequence using string search (Knuth et al. 1977) in combination with a trie structure; the position in a token line is found using hashing.

14.8 Extending PredC to Possible Worlds

To accommodate the classical modalities of necessity ($\Box$) and possibility ($\Diamond$) as well as change in time and space, Montague extended models like 14.6.3 into *model structures* by adding the infinite sets I for moments of time and J for possible worlds:

14.8.1 DEFINITION OF A MODEL STRUCTURE

A model structure is defined as the sextuple $<A, I, J, B, F, g>$, where A is an infinite set of objects or individuals, I an infinite set of moments of time, J an infinite set of possible worlds, B a finite set of basic expressions, F a denotation function from B into A^*, and g an assignment function from variables into A^*. The elements of J are in linear order and $I \times J$ is the Cartesian product of I and J.

The operators of 14.1.1 (3, 4) are extended to $\Box$ (necessary) and $\Diamond$ (possible):

Syntax:
(i) If ϕ is a sentence, then $\Box\phi$ is a sentence.
(j) If ϕ is a sentence, then $\Diamond\phi$ is a sentence.

Semantics:
(i) If ϕ is a sentence, then $\Box\phi^{\mathcal{M}, i\,j\,g}$ is 1 iff $\phi^{\mathcal{M}, i', j', g}$ is 1 for all $<i', j'> \varepsilon\ I \times J$.
(j) If ϕ is a sentence, then $\Diamond\phi^{\mathcal{M}, i\,j\,g}$ is 1 iff $\phi^{\mathcal{M}, i', j', g}$ is 1 for at least one $<i', j'> \varepsilon\ I \times J$.

This construct of possible world semantics raises the question: How can truth values characterize content? Let us consider an analogy from technology. On the computer screen, a black and white portrait is composed of two kinds of pixels, black and white, yet the pixel arrangement can be recognized as an individual face, and the more pixels the sharper the image. This is similar in a formal model with propositions which have only two denotations, $\emptyset$ and $\{\emptyset\}$.

Next compare a photo and a movie, both in black and white: they differ in that the pixels in a photo are static, as in the formal model 14.6.3, while the pixels in

---

[6] Brooks (1986): "The world is its own best model."

the movie change in time and location, as in the formal model structure 14.8.1. Like models, model structures rely on definitions produced by hand, resembling animation in film. TV and DBS, in contrast, may create content automatically, one by recording the agent-external reality, the other by monitoring cognition-external data.

TV and DBS differ in that (i) TV is limited to the output modalities of visual and auditory display, while DBS integrates a wide range of modalities in recognition as well as action, and (ii) TV is limited to displaying recorded images and sound, while DBS spontaneously processes (a) fresh input of raw data directly into content (recognition) and (b) currently activated content directly into raw data output (action), thus accommodating Hume's famous argument against postulating a *homunculus*[7] (Hume 1748). The technical basis is a content-addressable memory with a now front, clearance of the now front in functional intervals, automatic word form recognition and production, as well as operations for connecting input into content and for activating content by navigation, which are all absent in TV.

14.9 Semantic Relations of Structure

The proplets of (i) an elementary proposition and (ii) between the top verb proplets of coordinated propositions (TExer 2.1.6 ff.) as well as subordinated propositions (TExer Sects. 2.5, 2.6, 3.4–3.6) are connected by the four semantic relations of structure:

14.9.1 THE FOUR SEMANTIC RELATIONS OF STRUCTURE

1. subject/predicate
2. object\predicate
3. modifier|modified
4. conjunct−conjunct

As an intrapropositional example, consider the following content as a set of proplets:

14.9.2 CONTENT OF Lucy found a big blue square . AS PROPLET SET

⌈sur: lucy ⌉	⌈sur: ⌉	⌈sur: ⌉	⌈sur: ⌉	⌈sur: ⌉
noun: [person x]	**verb: find**	adj: **big**	adj: **blue**	**noun: square**
cat: snp	cat: n′ a′ decl	cat: adn	cat: adnv	cat: sn
sem: nm f	sem: ind past	sem: pad	sem: pad	sem: sg
fnc: find	**arg: [person x] square**	mdd: **square**	mdd:	**fnc: find**
mdr:	mdr:	mdr:	mdr:	mdr: **big**
nc:	nc:	nc: **blue**	nc:	nc:
pc:	pc:	pc:	pc: **big**	pc:
⌊prn: 14 ⌋	⌊prn: 14 ⌋	⌊prn: 14 ⌋	⌊prn: 14 ⌋	⌊prn: 14 ⌋

The subject/predicate relation is coded bidirectionally between (i) the core feature [noun: [person x]] of *lucy* and the continuation feature [arg: [person x] square]

---

[7] Rejecting a humunculus does not imply rejecting iconicity (similarity based on pattern matching), which is essential for computational similarity in DBS cognition (FoCL 3.3), pace Ogden&Richards (1923).

of *find* and (ii) the core feature [verb: find] of *find* and the continuation feature [fnc: find] of *lucy*. Similarly, the object\predicate relation is coded between (i) the core feature [noun: square] of *square* and the continuation feature [arg: [person x] square] of *find* and (ii) the core feature of [verb: find] of *find* and the continuation feature [fnc: find] of *square*. Correspondingly for *big* and *blue*.

By representing the semantic relations of structure with the operators /, \, |, and −, the content 14.9.2 may be shown graphically as follows:

14.9.3 SEMANTIC RELATIONS OF STRUCTURE IN THE CONTENT 14.9.2

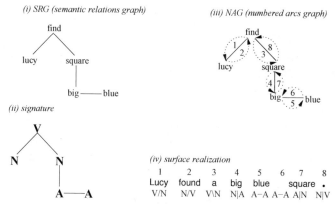

The graph is based on the proplets of 14.9.2. It characterizes the semantic relations of structure with four views. View (i), called the *semantic relations graph* (SRG), uses the core values lucy, find, square, big and blue as nodes. View (ii), called the *signature*, uses the core attributes N for noun, V for verb, and A for adj as nodes. View (iii), called the *numbered arcs graph* (NAG), supplements the SRG with numbered arcs which are used in (iv), called the *surface realization*, to show the navigation which activates content in the think mode and optionally realizes the language-dependent surfaces in a speak mode which rides piggyback on the think mode navigation. The concepts, as the elementary building blocks of DBS, are shown by placeholder values, using English base forms for convenience.

The operators /, \, |, and − are used also for extrapropositional relations:

14.9.4 RELATING TWO TRANSITIVE VERBS EXTRAPROPOSITIONALLY

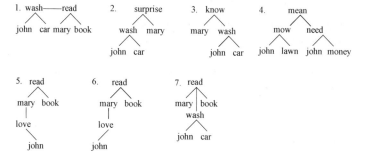

1. conjunct—conjunct: John washed the car. Mary read a book.

2. subject/predicate: That John washed the car surprised Mary.

3. object\predicate: Mary knew that John had washed the car.

4. sbj/prd\obj: That John mows the lawn means that John needs money.[8]

5. adn mdr|mdd (subject gap): Mary who loves John read a book.

6. adn mdr|mdd (object gap): Mary who(m) John loves read a book.

7. adv modifier|modified: When John washed the car Mary read a book.

Which kind of relation may connect the predicates of two component propositions depends on the verb class. For example, 2 requires a psych verb (TExer 2.5), while 3 requires a mental state verb (TExer 2.6). Thus, it is impossible linguistically to use the same predicate[9] for constructing all seven constellations.

Once the semantic structure of a grammatical construction has been figured out graphically, writing the content as a set of proplets, e.g., 14.9.2, is easy. The subsequent steps are (a) the hear mode derivation using the lexical proplets provided by automatic word form recognition and (b) the think-speak mode derivation using the content derived in the hear mode (laboratory set-up). For details see TExer.

14.10 Properties Common to Hear, Think, and Think-Speak Operations

As an agent-based data-driven approach, DBS derivations require three kinds of operations: for (i) the hear mode, (ii) the think mode, and (iii) the think-speak mode. They have in common that they use a time-linear derivation order and share the following structural properties:

14.10.1 STRUCTURAL PROPERTIES COMMON TO ALL DBS OPERATIONS

1. DBS operations consist of an antecedent, a connective, and a consequent.

2. The antecedent and the consequent are defined as sets of proplet patterns which are semantically connected by proplet-internal addresses.[10].

3. With the exception of inferences, the proplet patterns of the antecedent are the input pattern, while the proplet patterns of the consequent are the output pattern.

4. Inferences apply by matching input to the antecedent (deductive use, CC 3.5.1) or the consequent (abductive use, CC 3.5.2).

---

[8] Verbs which take a clausal subject and a clausal object simultaneously (class 4 in 14.9.4) include *entail, hint, imply, indicate, mean, presuppose,* and *suggest.* Thanks to Prof. Kiyong Lee for pointing it out.

[9] The criteria for establishing word classes are diverse (Levin 2009). For analyzing semantic relations of structure, DBS defines verb classes in terms of their possible valency filler (CLaTR 15).

[10] With the exception of suspension operations (14.11.1, 3) in the hear mode.

5. An operation is activated by content matching the input pattern (data-driven).

6. Binding the input constants to the variables of the input pattern enables the output pattern to derive the output.

7. The codomain of a variable in a pattern proplet may be restricted by an explicit list of possible values (variable restriction, e.g., CC 5.5.4).

The hear mode is the language variant of the agent's recognition. Language recognition, regardless of the medium (e.g., audition, vision) differs from nonlanguage recognition in that it provides an explicit processing order in the form of a left-associative surface sequence as input. The think mode controls the agent's actions. The operations of the think mode are of two kinds, (i) activation by navigation and (ii) inferencing. In the think-speak mode, either may be realized in the agent's natural language, technically based on lexicalization rules which sit in the sur slot of think mode operation patterns.

The application of hear, think, and think-speak mode operations in DBS is driven by computational pattern matching. Therefore truly efficient pattern matching is of the essence. It is easily achieved by definition (i) fixing the attributes in matching pattern and input proplets to the same order, and (ii) by excluding recursion.[11]

14.11   Hear Mode Operations

For building complex content in the hear mode of natural language communication, a next word (i.e., a lexical proplet provided by automatic word form recognition) is connected to the current sentence start by one of the following operation kinds:

14.11.1 THREE KINDS OF HEAR MODE OPERATIONS

1. cross-copying (connective $\times$, as in SBJ$\times$PRD; TExer 6.3.1, 1)

2. absorption (connective $\cup$, as in DET$\cup$CN; TExer 6.3.1, 51)

3. suspension (connective $\sim$, as in ADV$\sim$NOM; TExer 6.3.1, 32)

A hear mode operation takes two proplets as input and produces one or two proplets as output. The pattern matching is controlled by the syntactic category, regardless of the distinction between the semantic kinds concept, indexical, and name.

The following hear mode derivation of the content 14.9.2 uses operations of the kind (i) cross-copying and (ii) absorption. For (iii) suspension see TExer, 3.1.

---

[11] Motivated by an inappropriate notion of generality, order-free feature structures with recursion are today's standard (Carpenter 1992), but uniquely ill-suited for efficient pattern matching.

14.11.2 TIME-LINEAR SURFACE-COMPOSITIONAL DERIVATION

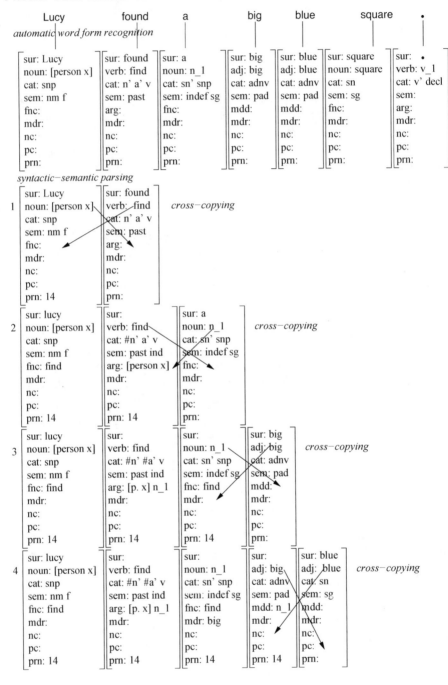

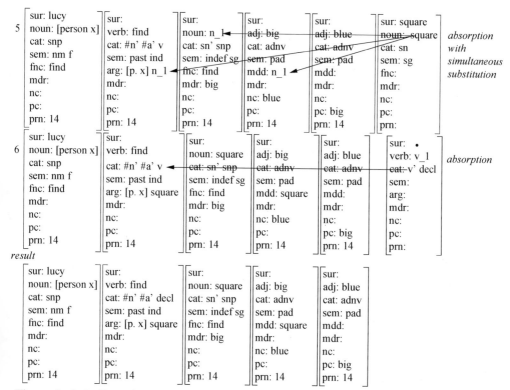

The analysis is (i) *surface-compositional* because each lexical item has a concrete sur value and there are no surfaces without a proplet analysis. The derivation order is (ii) *time-linear*, as shown by the stair-like addition of a next word proplet. The activation and application of operations is (iii) *data-driven* by automatic word form recognition.

The computational pattern matching of DBS operations is illustrated by the following application of a hear mode operation:

14.11.3 CROSS-COPYING *lucy* AND *find* WITH SBJ×PRD (line 1 in 19.3.3)

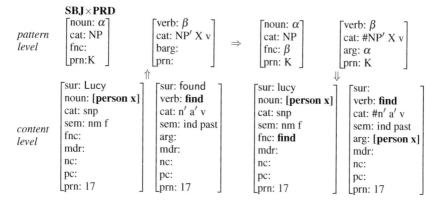

Lexical lookup and syntactic-semantic concatenation are incrementally intertwined: the lookup of a new next word occurs only after the current next word has been processed into the current sentence start. In each concatenation, the language-dependent sur value provided by lexical lookup is omitted in the output, with the partial exception of names.

While a cross-copying operation like 14.11.3 produces two output proplets, an absorption operation produces only one. Consider the following example:

14.11.4 ABSORBING INTERPUNCTUATION INTO *find* WITH S∪IP (line 6)

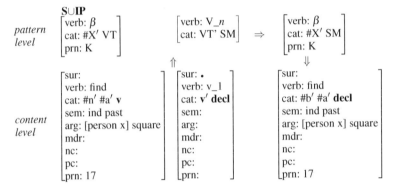

The names of all three kinds of DBS hear mode operations (14.11.1) have an input1 connective input2 structure, where input2 matches a next word and input1 looks for matching input in the sentence start, i.e., the set of proplets at the current now front.

14.12 Activation in the Think and Think-Speak Modes

Of the two kinds of DBS think mode operations, navigation operations serve to activate existing content and consist of one input and one output pattern. Inference operations serve to derive new content from given content and consist of an open number of input and output patterns. Both kinds use a time-linear derivation order and both may optionally produce language-dependent surfaces relying on the same lex rules.

Intrapropositionally, navigational DBS think and think-speak mode operations alike activate content by traversing the semantic relations of structure in both directions. Accordingly, the intrapropositional operations traversing subject/predicate are V/N and N/V, those traversing object\predicate are V\N and N\V, those traversing the modifier|modified relation adnominally are N↓A and A↑N (and similarly for adverbial V|A), and those traversing the conjunct−conjunct relation N−N are N→N and N←N (and similarly for A−A and V−V).[12]

The same holds for extrapropositional activations, except that extrapropositional coordination is uni-directional in the direction of time and requires an inference for

---
[12] For a more detailed account see TExer.

traversal in the anti-temporal direction. In the DBS graph analysis 14.9.3, which happens to be intrapropositional, the dual traversals are shown in the (iii) NAG (numbered arcs graph) and applied in the (iv) surface realization.

The following examples refer to the arc numbers in the NAG of 14.9.3:

14.12.1 NAVIGATING WITH V⸝N FROM *find* TO *Lucy* (arc 1)

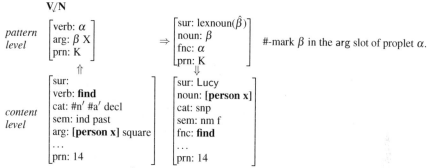

In the think mode, the lexnoun($\hat{\beta}$) operation in the sur slot of the output proplet is switched off, but switched on in the think-speak mode. The variable $\hat{\beta}$ refers to a list which associates each core value with a language-dependent counterpart – with the exception of names, which are realized from a marker in the sur slot (CASM'17).

14.12.2 NAVIGATING WITH N⸝V FROM *Lucy* BACK TO *find* (arc 2)

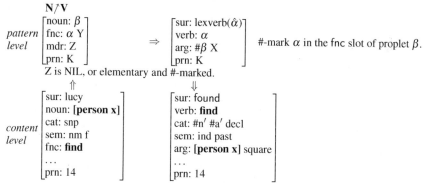

Production of the surface found by lexverb($\hat{\alpha}$) is based on the features [verb: find], [cat: #n′ decl], and [sem: ind past] of the output proplet.

14.13 Inferencing

For inferencing as the third main part of cognition, besides recognition and action, please see CC, 3.4-3.6, 4, 5, and Part II.

14.14   Conclusion

The main difference between the semantics of DBS and PredC is the agent-based data-driven ontology of DBS vs. the sign-based substitution-driven ontology of PredC. As a consequence, DBS requires an on-board interface component and an on-board memory, while PredC does not. Also, a sign-based approach has no room for distinct hear and speak modes, while DBS treats them as the language part of recognition and action.

PredC is perhaps more general and parsimonious, but DBS avoids some longstanding problems of PredC, such as over- (14.2) and undergeneration (14.4), and a dependence of contingent meaning on hand-crafted models (14.6 and 14.8). Also, because DBS implements agent-based recognition and action, including the hear (14.11), think, and think-speak modes (14.12), it is better suited for building a talking autonomous robot than systems inherently without on-board interface and memory components.

15. Agent-Based Memory as an On-Board Database

Today's databases are computer memories used for storage in, and retrieval from, large collections of systematic information, such as bibliographical, medical, trade, banking, insurance, and tax data. In natural agents, cognitive content is stored in, and retrieved from, the brain's memory. In artificial DBS agents, natural memory is reconstructed as an on-board database.

An agent-based database serving as the memory of an artificial agent requires interaction with (a) an interface component for automatic recognition and action, and (b) a data-driven operations component. Basic tasks are (i) transfer of content from a speaker to a hearer in natural language communication (15.1–15.5) (ii) coactivation of memory content by current processing (15.6–15.8), and (iii) an algorithm for efficient non-language recognition, e.g., vision (15.9), and action, e.g., manipulation. The goal is *functional equivalence* between the natural prototype and its computational reconstruction at appropriate levels of abstraction.

15.1 Input-Output of Conventional Database vs. On-Board Memory

A conventional database has, simply put, the (i) input constellation of a programmer storing data and the (ii) output constellation of a user retrieving copies of the data:

15.1.1 CONVENTIONAL DATABASE INTERACTION

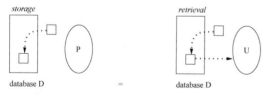

Interaction takes place between different agents, i.e., the programmer (oval P) and the user (oval U), and a single database, i.e., the large box D. The small boxes represent the content serving as input and output. Agent P controls the storage and agent U controls the retrieval operations. Both use the same database and the same programming language, e.g., SQL (structured query language), the commands of which are executed as electronic procedures.

Next consider the transfer of content from speaker to hearer by means of raw data:

15.1.2 SPEAKER AND HEARER INTERACTING IN COMMUNICATION

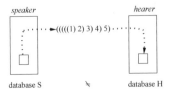

The speaker and the hearer use different on-board databases, S and H, which may be natural or artificial and typically contain different contents. In natural language communication, agents alternate between the speak and the hear mode (turn-taking). In the speak mode, automatic word form production takes cognitive content as input and maps it into language-dependent surfaces as agent-external raw data output. In the hear mode, automatic word form recognition takes agent-external raw data as input and maps it into cognitive content as output.

The raw data transporting content from the speaker to the hearer have the left-associative (time-linear) structure[1] shown in 15.1.2 abstractly as $(((((1) 2) 3) 4) 5)$. Otherwise the agent-external raw surface data have neither meaning nor any grammatical properties whatsoever (no reification in DBS), but may be measured by the natural sciences. For word form recognition and production, DBS uses computational pattern matching based on the type-token relation from philosophy (Peirce 1906, CP Vol. 4, p. 375).

15.2 Data Structure and Operations in a Record-Based Database

A computational database is defined by (i) a database schema, (ii) a data structure, and (iii) an algorithm for the storage, retrieval, and processing of content. The databases used most widely in business are the relational databases with the data structure of *records*:

15.2.1 RECORDS OF A RELATIONAL DATABASE

	last name	first name	place	...
A1	Schmidt	Peter	Bamberg	...
A2	Meyer	Susanne	Nürnberg	...
A3	Sanders	Reinhard	Schwabach	...
	⋮	⋮	⋮	

The columns, named by different attributes like first name, last name, etc., are called the fields of the record type. The lines A1, A2, etc., each constitute a record.

---

[1] Aho and Ullman (1977, p. 47)

The data structure determines the operations of a database. The standard operations of a record-based database are the storage, retrieval, update, and recombining of information. For example, to retrieve the name of the representative in Schwabach, the user types the following SQL command:

15.2.2 DATABASE QUERY

Query:
```
select A#
where city = 'Schwabach'
```
Result:
```
result: A3 Sanders Reinhard
```

Other SQL commands are outer join and inner join for the conjunction and intersection of record columns. Change of data is carried out by replacement operations. Illegal modification as in theft is prevented by differentiating access privileges.

In summary, record-based databases are agent-driven: human commands initiate quasi-mechanical procedures which correspond to storing, retrieving, re-sorting, and correcting cards in a filing cabinet. Compared to the nonelectronic method, the computational system has practical advantages. The adding, finding, recombining, and correcting of information is largely automated, making it faster, and the possibilities of search are more powerful because the records of different fields may be set-theoretically manipulated for complex queries.

15.3 Data Structure and Operations

Alternative to records, the data structure of DBS is non-recursive feature structures with ordered attributes called *proplets*, connected into content by the four semantic relations of structure, i.e., subject/predicate, object\predicate, modifier|modified, and conjunct–conjunct, coded by address. Consider the following representation of the record A3 in 15.2.1 as a set of proplets:

15.3.1 THE CONTENT OF "R.S. resides in S." IN DBS

$$
\begin{bmatrix}
\text{sur: reinhard sanders} \\
\text{noun: [\textbf{person x}]} \\
\text{cat: snp} \\
\text{sem: nm m sg} \\
\text{fnc: \textbf{reside}} \\
\text{mdr:} \\
\text{nc:} \\
\text{pc:} \\
\text{prn: 26}
\end{bmatrix}
\begin{bmatrix}
\text{sur:} \\
\text{verb: \textbf{reside}} \\
\text{cat: \#n' \#mdr' decl} \\
\text{sem:} \\
\text{arg: [\textbf{person x}]} \\
\text{mdr: [\textbf{town y}]} \\
\text{nc:} \\
\text{pc:} \\
\text{prn: 26}
\end{bmatrix}
\begin{bmatrix}
\text{sur: schwabach} \\
\text{noun: [\textbf{town y}]} \\
\text{cat: snp} \\
\text{sem: \textit{in} nm sg} \\
\text{mdd: \textbf{reside}} \\
\text{mdr:} \\
\text{nc:} \\
\text{pc:} \\
\text{prn: 26}
\end{bmatrix}
$$

The subject/predicate relation is coded by [fnc: reside] in the [*person x*] proplet and [arg: [person x]] in the *reside* proplet. The modifier|modified relation is coded by [mdd: reside] in the [*town y*] proplet and [mdr: [town y]] in the *reside* proplet.

The attributes of the proplets are (1) the sur attribute for the optional language-dependent surface, (2) the core attribute for the obligatory values of noun, verb, or adj (functor-argument), (3) the cat(egory) and (4) sem(antics) attributes for syntactic-semantic information, the (5, 6) attributes fnc, arg, mdr, and mdd for functor-argument continuation,[2] the attributes (7) nc (next conjunct) and (8) pc (previous conjunct)[3] coordination continuation, and (9) the prn attribute for the proposition number.

As a set, the proplets of a content are order-free, which is essential for storage in and retrieval from a content-addressable database (Chisvin and Duckworth 1992). Instead of sorting data into records like name, place, customer, or employee, content-addressable DBS stores proplets based on the letter sequence of their core value, enabling computational string search. The database schema of DBS consists (i) horizontally of proplets with the same core value in the time-linear order of arrival, called *token lines*, and (ii) vertically in a *column of token lines* in the alphabetical order induced by the core values. The key is defined as the proplet address, consisting of the core and the prn value, e.g., [reside 26].:

15.3.2 SORTING THE PROPLETS OF 15.3.1 INTO DBS DATABASE SCHEMA

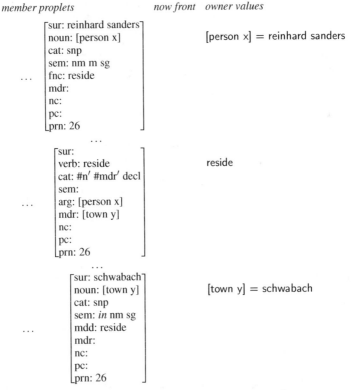

The number and the length of token lines are unrestricted.[4]

---

[2] The fnc-arg values are obligatory and the mdr-mdd values are optional.

15.4 The On-Board Orientation System (OBOS)

The change from a sign-based substitution-driven to an agent-based data-driven ontology lead to two fundamental innovations: (i) the *on-board orientation system* (OBOS) and (ii) the *now front*. Both are part of the agent's cognition, but the OBOS complements the DBS database from the outside, while the now front is the processing arena inside the on-board database.

One purpose of the OBOS is the interpretation of indexicals, the other the type-token distinction and with it the distinction between semantics and pragmatics. The input to the OBOS are parameter values provided by the interface component. The output is a continuous monitoring in form of the STAR proplet, named after the attributes Space, Time, Agent (speaker), and Recipient (intended hearer).[5]

15.4.1 EXAMPLE OF A STAR PROPLET

$$
\begin{bmatrix}
\text{S: yard} \\
\text{T: 2007-04-05T14:30} \\
\text{A: sylvester} \\
\text{R: speedy} \\
\text{3rd:} \\
\text{prn: 39}
\end{bmatrix}
$$

The S and the T values are the agent's current location and moment of time. The A, R, and 3rd values refer to speaker, hearer, and 3rd person pronoun. The A value is constant, while the other values are provided by continuous monitoring of the OBOS.

A proposition without a STAR is a content type and represents the semantics:

15.4.2 TYPE OF THE NONLANGUAGE CONTENT I saw you

In a type, the prn values are substitution variables, here K. Without a STAR, the indexicals, here pro1 (I) and pro2 (you), are left 'dangling.'

A proposition connected to a STAR, in contrast, is a content token (pragmatics):

---

[3] Prof. Meyer-Wegener, Chair of Databases at Erlangen Computer Science, pointed out that the DBS solution can be simulated by the record-based approach. As proof, the Java implementation of DBS by Arcadius Kycia (Kycia 2004), student at Erlangen Computerlinguistics, was reprogrammed by Wolfgang Fischer, student at Erlangen Informatik, as an RDBMS (Fischer 2002). To enable the comparison, the different input-output constellations of DBS (15.1.2) and RDBMS (15.1.1) were left aside by using the laboratory set-up (TExer 1.5). The RDBMS reconstruction succeeded, but turned out to be impractical for systematic upscaling.

[4] Backward navigation of an extrapropositional coordination is by an inference which provides appropriate conjunctions such as before that or three days earlier.

[5] For pronunciation, the STAR acronym omits the attributes 3rd and prn.

15.4.3 TOKEN OF THE NONLANGUAGE CONTENT I saw you

$$
\begin{bmatrix}
\text{sur:} \\
\text{noun: pro1} \\
\text{cat: s1} \\
\text{sem: sg} \\
\text{fnc: see} \\
\text{mdr:} \\
\text{nc:} \\
\text{pc:} \\
\text{prn: 12}
\end{bmatrix}
\begin{bmatrix}
\text{sur:} \\
\text{verb: see} \\
\text{cat: } \#n' \ \#a' \ \text{decl} \\
\text{sem: past ind} \\
\text{arg: pro1 pro2} \\
\text{mdr:} \\
\text{nc:} \\
\text{pc:} \\
\text{prn: 12}
\end{bmatrix}
\begin{bmatrix}
\text{sur:} \\
\text{noun: pro2} \\
\text{cat: sp2} \\
\text{sem:} \\
\text{fnc: see} \\
\text{mdr:} \\
\text{nc:} \\
\text{pc:} \\
\text{prn: 12}
\end{bmatrix}
\overset{\textit{STAR-0 proplet of origin}}{
\begin{bmatrix}
\text{S: yard} \\
\text{T: thursday} \\
\text{A: sylvester} \\
\text{R: hector} \\
\text{3rd:} \\
\text{prn: 12}
\end{bmatrix}
}
$$

The connection between a proposition and a STAR is a shared prn value, here 12. The indexical pro1 points at the A value sylvester, the indexical pro2 at the R value hector, and the indexical past at the present time value T, here thursday.

15.5 Loom-Like Clearance of the Now Front

The 2nd innovation of the DBS database schema is the *now front*[6]. Before processing a next proposition, the now front is cleared:[7]

15.5.1 TOKEN LINE WITH CLEAR NOW FRONT

| (i) member proplets | | | (ii) now front | (iii) owner |

$$
\overset{\textit{(i) member proplets}}{
\begin{bmatrix}
\text{noun: square} \\
\text{mdr:} \\
\text{nc:} \\
\text{pc:} \\
\text{prn: 3}
\end{bmatrix}
\begin{bmatrix}
\text{noun: square} \\
\text{mdr:} \\
\text{nc:} \\
\text{pc:} \\
\text{prn: 6}
\end{bmatrix}
}
\overset{}{
\begin{bmatrix}
\text{noun: square} \\
\text{mdr:} \\
\text{nc:} \\
\text{pc:} \\
\text{prn: 14}
\end{bmatrix}
}
\qquad \overset{\textit{(iii) owner}}{\text{square}}
$$

Clearance consists in moving the now front with the owners into fresh memory space (*loom-like clearance*), leaving the concatenated proplets behind in the field of member proplets, never to be changed, like *sediment*. Correcting content is limited to *adding* content, as in a diary entry, using reference by address.

For storage, a proplet provided by automatic word form recognition is written to the now front in the token line of the owner[8] which equals its core value. For declarative retrieval (in contrast to retrieval by pointer), the first step is going to the owner corresponding to the sought proplet's core value (vertical) and the second step is going along the token line to the sought proplet's prn value (horizontal).

---

[6] The now front made it possible to replace the rule packages of earlier LAG by data-driven application: in the hear mode, a next word proplet (i) is stored in its token line at the now front prior to processing and (ii) activates all operations matching it with their second input pattern; the activated operations (iii) look for a proplet at the now front matching their first input pattern, and (iv) apply if they find one (CC 2.2).

[7] For step by step derivations of now front states see CLaTR 13.3; NLC 11.2, 11.3. For derivations of the hear and the speak mode of 24 linguistically informed examples see TExer.

[8] The terminology of member proplets and owner values is reminiscent of the member and owner records in a classic network database (Elmasri and Navathe ([1989] 2017), which inspired the database schema of the A-memory in DBS.

The now front is cleared when its proplets have ceased to be candidates for further processing. This is basically the case when an elementary proposition is completed, formally indicated by the automatic incrementation of the prn value for the next proposition (NLC 13.5.1). Partial exceptions are the extrapropositional operations of (i) coordination (NLC 11) and (ii) functor-argument (NLC 7; TExer 2.5, 2.6, 3.3–3.5). In these cases, the verb of the completed proposition must remain at the now front for cross-copying with or navigation to the verb of the next proposition until the extrapropositional relation has been established or utilized.

Because the proplets at the current now front are limited to an elementary proposition, their number, vertically over the whole column, is usually no more than four or five. Horizontally, the number of proplets in a token line affected by a clearance is either zero or one.[9]

15.6 Resonating Content 1: Coactivation by Similarity

An important property of natural cognition is *association*, i.e. the automatic activation of content in memory which is related to content at the current now front. The associated content enriches the current content with individual reminiscences as well as general knowledge. In DBS, this process is modeled as the automatic *coactivation* of related contents stored in memory, called *resonating content*. The three method of coactivation are (1) similarity, (2) intersection, and (3) continuation.

Contents are computationally similar if they match the same pattern (CC 14.1). Degrees of similarity vary with the degree of pattern abstraction. Abstraction in proplets is systematically controlled by (i) the replacement of constants with variables and (ii) restrictions on the variables (CC 14.2). Patterns of increasing abstraction degrees coactivate sets of increasing size in which set n is contained in set n+1.

15.7 Resonating Content 2: Coactivation by Token Line Intersection

The second method of coactivating content is intersecting token lines to find instances of the same semantic relation between similar proplets in different propositions. 2nd degree search patterns derive from intersecting two proplets:

---

[9] Except for those rare cases in which several proplets have the same core and prn value, as in Oh Mary, Mary, Mary! or slept and slept and slept. When a now front slot is filled, a new free slot is opened. Thus, the token line of *sleep* would contain three instances of *sleep* proplets with the same prn value at the now front. They will all be left behind during the next loom-like clearance.

15.7.1 DERIVATION OF TWO INTERSECTING SEARCH PATTERNS

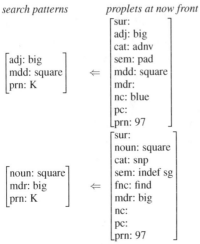

search patterns   *proplets at now front*

$$
\begin{bmatrix} \text{adj: big} \\ \text{mdd: square} \\ \text{prn: K} \end{bmatrix} \Longleftarrow \begin{bmatrix} \text{sur:} \\ \text{adj: big} \\ \text{cat: adnv} \\ \text{sem: pad} \\ \text{mdd: square} \\ \text{mdr:} \\ \text{nc: blue} \\ \text{pc:} \\ \text{prn: 97} \end{bmatrix}
$$

$$
\begin{bmatrix} \text{noun: square} \\ \text{mdr: big} \\ \text{prn: K} \end{bmatrix} \Longleftarrow \begin{bmatrix} \text{sur:} \\ \text{noun: square} \\ \text{cat: snp} \\ \text{sem: indef sg} \\ \text{fnc: find} \\ \text{mdr: big} \\ \text{nc:} \\ \text{pc:} \\ \text{prn: 97} \end{bmatrix}
$$

Automatically derived from concept proplets at the current now front, intersecting search patterns are moved along their token lines from right to left (backwards in time).

The search patterns in 15.7.1 express the modifier|modified relation between big and square. As they are moved in parallel along their token lines (NLC 5.1) in a DBS database, they retrieve pairs of proplets connected by (i) a shared prn value and (ii) the same semantic relation, as in the following example:

15.7.2 2ND DEGREE INTERSECTION COACTIVATING big square

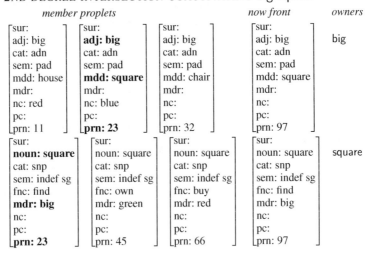

member proplets   *now front*   *owners*

$$
\begin{bmatrix} \text{sur:} \\ \text{adj: big} \\ \text{cat: adn} \\ \text{sem: pad} \\ \text{mdd: house} \\ \text{mdr:} \\ \text{nc: red} \\ \text{pc:} \\ \text{prn: 11} \end{bmatrix}
\begin{bmatrix} \text{sur:} \\ \textbf{adj: big} \\ \text{cat: adn} \\ \text{sem: pad} \\ \textbf{mdd: square} \\ \text{mdr:} \\ \text{nc: blue} \\ \text{pc:} \\ \textbf{prn: 23} \end{bmatrix}
\begin{bmatrix} \text{sur:} \\ \text{adj: big} \\ \text{cat: adn} \\ \text{sem: pad} \\ \text{mdd: chair} \\ \text{mdr:} \\ \text{nc:} \\ \text{pc:} \\ \text{prn: 32} \end{bmatrix}
\begin{bmatrix} \text{sur:} \\ \text{adj: big} \\ \text{cat: adn} \\ \text{sem: pad} \\ \text{mdd: square} \\ \text{mdr:} \\ \text{nc:} \\ \text{pc:} \\ \text{prn: 97} \end{bmatrix} \text{big}
$$

$$
\begin{bmatrix} \text{sur:} \\ \textbf{noun: square} \\ \text{cat: snp} \\ \text{sem: indef sg} \\ \text{fnc: find} \\ \textbf{mdr: big} \\ \text{nc:} \\ \text{pc:} \\ \textbf{prn: 23} \end{bmatrix}
\begin{bmatrix} \text{sur:} \\ \text{noun: square} \\ \text{cat: snp} \\ \text{sem: indef sg} \\ \text{fnc: own} \\ \text{mdr: green} \\ \text{nc:} \\ \text{pc:} \\ \text{prn: 45} \end{bmatrix}
\begin{bmatrix} \text{sur:} \\ \text{noun: square} \\ \text{cat: snp} \\ \text{sem: indef sg} \\ \text{fnc: buy} \\ \text{mdr: red} \\ \text{nc:} \\ \text{pc:} \\ \text{prn: 66} \end{bmatrix}
\begin{bmatrix} \text{sur:} \\ \text{noun: square} \\ \text{cat: snp} \\ \text{sem: indef sg} \\ \text{fnc: find} \\ \text{mdr: big} \\ \text{nc:} \\ \text{pc:} \\ \text{prn: 97} \end{bmatrix} \text{square}
$$

The big|square relation retrieved has the prn value 23, in contrast to the prn value 97 of the trigger at the now front.

An intersection of two token lines is of degree 2, of three token lines of degree 3, and so on. The following example derives a search patterns for a 3rd degree intersection:

15.7.3 SEARCH PATTERNS FOR A 3RD DEGREE INTERSECTION

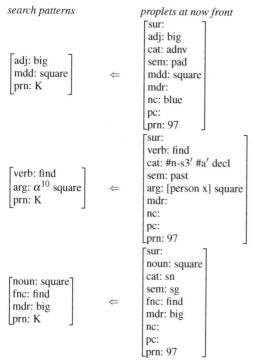

search patterns   *proplets at now front*

$$\begin{bmatrix} \text{adj: big} \\ \text{mdd: square} \\ \text{prn: K} \end{bmatrix} \Leftarrow \begin{bmatrix} \text{sur:} \\ \text{adj: big} \\ \text{cat: adnv} \\ \text{sem: pad} \\ \text{mdd: square} \\ \text{mdr:} \\ \text{nc: blue} \\ \text{pc:} \\ \text{prn: 97} \end{bmatrix}$$

$$\begin{bmatrix} \text{verb: find} \\ \text{arg: } \alpha^{10} \text{ square} \\ \text{prn: K} \end{bmatrix} \Leftarrow \begin{bmatrix} \text{sur:} \\ \text{verb: find} \\ \text{cat: \#n-s3' \#a' decl} \\ \text{sem: past} \\ \text{arg: [person x] square} \\ \text{mdr:} \\ \text{nc:} \\ \text{pc:} \\ \text{prn: 97} \end{bmatrix}$$

$$\begin{bmatrix} \text{noun: square} \\ \text{fnc: find} \\ \text{mdr: big} \\ \text{prn: K} \end{bmatrix} \Leftarrow \begin{bmatrix} \text{sur:} \\ \text{noun: square} \\ \text{cat: sn} \\ \text{sem: sg} \\ \text{fnc: find} \\ \text{mdr: big} \\ \text{nc:} \\ \text{pc:} \\ \text{prn: 97} \end{bmatrix}$$

For illustrating a 3rd degree search with the 15.7.3 patterns, the content in the DBS database sketch 15.7.2 is assumed to be extended:

15.7.4 RESONATING CONTENT RESULTING FROM A 3RD DEGREE SEARCH

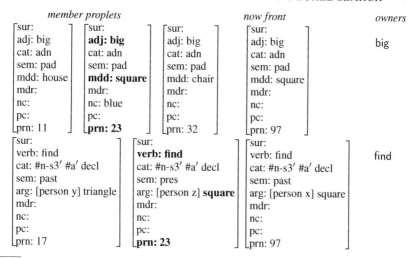

member proplets   *now front*   *owners*

$$\begin{bmatrix} \text{sur:} \\ \text{adj: big} \\ \text{cat: adn} \\ \text{sem: pad} \\ \text{mdd: house} \\ \text{mdr:} \\ \text{nc:} \\ \text{pc:} \\ \text{prn: 11} \end{bmatrix} \begin{bmatrix} \text{sur:} \\ \textbf{adj: big} \\ \text{cat: adn} \\ \text{sem: pad} \\ \textbf{mdd: square} \\ \text{mdr:} \\ \text{nc: blue} \\ \text{pc:} \\ \textbf{prn: 23} \end{bmatrix} \begin{bmatrix} \text{sur:} \\ \text{adj: big} \\ \text{cat: adn} \\ \text{sem: pad} \\ \text{mdd: chair} \\ \text{mdr:} \\ \text{nc:} \\ \text{pc:} \\ \text{prn: 32} \end{bmatrix} \quad \begin{bmatrix} \text{sur:} \\ \text{adj: big} \\ \text{cat: adn} \\ \text{sem: pad} \\ \text{mdd: square} \\ \text{mdr:} \\ \text{nc:} \\ \text{pc:} \\ \text{prn: 97} \end{bmatrix} \quad \text{big}$$

$$\begin{bmatrix} \text{sur:} \\ \text{verb: find} \\ \text{cat: \#n-s3' \#a' decl} \\ \text{sem: past} \\ \text{arg: [person y] triangle} \\ \text{mdr:} \\ \text{nc:} \\ \text{pc:} \\ \text{prn: 17} \end{bmatrix} \begin{bmatrix} \text{sur:} \\ \textbf{verb: find} \\ \text{cat: \#n-s3' \#a' decl} \\ \text{sem: pres} \\ \text{arg: [person z] } \textbf{square} \\ \text{mdr:} \\ \text{nc:} \\ \text{pc:} \\ \textbf{prn: 23} \end{bmatrix} \quad \begin{bmatrix} \text{sur:} \\ \text{verb: find} \\ \text{cat: \#n-s3' \#a' decl} \\ \text{sem: past} \\ \text{arg: [person x] square} \\ \text{mdr:} \\ \text{nc:} \\ \text{pc:} \\ \text{prn: 97} \end{bmatrix} \quad \text{find}$$

---

[10] See CC 13 for more on the coactivation of content by autonomous navigation.

$$
\begin{bmatrix}
\text{sur:} \\
\textbf{noun: square} \\
\text{cat: snp} \\
\text{sem: indef sg} \\
\text{fnc: find} \\
\textbf{mdr: big} \\
\text{nc:} \\
\text{pc:} \\
\textbf{prn: 23}
\end{bmatrix}
\begin{bmatrix}
\text{sur:} \\
\text{noun: square} \\
\text{cat: snp} \\
\text{sem: indef sg} \\
\text{fnc: own} \\
\text{mdr: green} \\
\text{nc:} \\
\text{pc:} \\
\text{prn: 45}
\end{bmatrix}
\begin{bmatrix}
\text{sur:} \\
\text{noun: square} \\
\text{cat: snp} \\
\text{sem: indef sg} \\
\text{fnc: buy} \\
\text{mdr: red} \\
\text{nc:} \\
\text{pc:} \\
\text{prn: 66}
\end{bmatrix}
\begin{bmatrix}
\text{sur:} \\
\text{noun: square} \\
\text{cat: snp} \\
\text{sem: indef sg} \\
\text{fnc: sell} \\
\text{mdr: big} \\
\text{nc:} \\
\text{pc:} \\
\text{prn: 97}
\end{bmatrix}
\quad \text{square}
$$

Compared to the number of proplets in complete token lines, the number of resonating proplets in an intersection is (i) greatly reduced and (ii) more precisely adapted to the triggers at the agent's current now front content. As the computational counterpart to associating freely in natural cognition, artificial coactivation is an important part of automated reasoning. For more precision, the number of intersections may be increased. For more generality, core values provided by the now front may be replaced with more general terms in the associated semantic field hierarchies.

15.8 Resonating Content 3: Coactivation by Continuation

The third method of coactivating content resembles Quillian's (1968) *spreading activation*, but is more constrained in that it is restricted to navigating along existing semantic relations between proplets in the artificial agent's on-board database. The method activates all proplets in memory which correspond to a single trigger at the current now front and follows their semantic relations intra- and extrapropositionally to 'explore the neighborhood'. This has the potential of providing cognition with relevant information for further reasoning. Like intersection, coactivation by continuation is based technically on the database schema and the data structure of DBS:

15.8.1 CONTENT SUPPORTING INTRAPROPOSITIONAL ACTIVATION

$$
\begin{bmatrix}
\text{sur: lucy} \\
\textbf{noun: [person x]} \\
\text{cat: snp} \\
\text{sem: nm f} \\
\textbf{fnc: find} \\
\text{mdr:} \\
\text{nc:} \\
\text{pc:} \\
\text{prn: 23}
\end{bmatrix}
\begin{bmatrix}
\text{sur:} \\
\textbf{verb: find} \\
\text{cat: \#n' \#a' decl} \\
\text{sem: pres} \\
\text{arg: \textbf{[person x] square}} \\
\text{mdr:} \\
\text{nc: (be 24)} \\
\text{pc:} \\
\text{prn: 23}
\end{bmatrix}
\begin{bmatrix}
\text{sur:} \\
\textbf{noun: square} \\
\text{cat: snp} \\
\text{sem: def sg} \\
\textbf{fnc: find} \\
\textbf{mdr: big} \\
\text{nc:} \\
\text{pc:} \\
\text{prn: 23}
\end{bmatrix}
\begin{bmatrix}
\text{sur:} \\
\textbf{adj: big} \\
\text{cat: adnv} \\
\text{sem: pad} \\
\textbf{mdd: square} \\
\text{mdr:} \\
\textbf{nc: blue} \\
\text{pc:} \\
\text{prn: 23}
\end{bmatrix}
\begin{bmatrix}
\text{sur:} \\
\textbf{adj: blue} \\
\text{cat: adnv} \\
\text{sem: pad} \\
\text{mdd:} \\
\text{mdr:} \\
\text{nc:} \\
\textbf{pc: big} \\
\text{prn: 23}
\end{bmatrix}
$$

The values in bold face are either core or continuation values and specify the semantic relations of structure between the proplets derived in CC 2.1.3.

Once a coactivation has resulted in the traversal of a proposition, it may continue to the next, as in Lucy found a big blue square. She was happy.:

15.8.2 Coactivation Moving from One Proposition to the Next

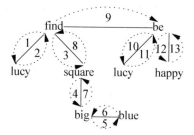

This content is defined as the following set of proplets connected by address:

15.8.3 Content Supporting Extrapropositional Coactivation

$$
\begin{bmatrix}
\text{sur: lucy} \\
\text{noun: [person x]} \\
\text{cat: snp} \\
\text{sem: nm f} \\
\text{fnc: find} \\
\text{mdr:} \\
\text{nc:} \\
\text{pc:} \\
\text{prn: 23}
\end{bmatrix}
\begin{bmatrix}
\text{sur:} \\
\text{verb: find} \\
\text{cat: \#n' \#a' decl} \\
\text{sem: pres} \\
\text{arg: [person x] square} \\
\text{mdr:} \\
\text{nc: (be 24)} \\
\text{pc:} \\
\text{prn: 23}
\end{bmatrix}
\begin{bmatrix}
\text{sur:} \\
\text{noun: square} \\
\text{cat: snp} \\
\text{sem: def sg} \\
\text{fnc: find} \\
\text{mdr: big} \\
\text{nc:} \\
\text{pc:} \\
\text{prn: 23}
\end{bmatrix}
\begin{bmatrix}
\text{sur:} \\
\text{adj: big} \\
\text{cat: adnv} \\
\text{sem: pad} \\
\text{mdd: square} \\
\text{mdr:} \\
\text{nc: blue} \\
\text{pc:} \\
\text{prn: 23}
\end{bmatrix}
\begin{bmatrix}
\text{sur:} \\
\text{adj: blue} \\
\text{cat: adnv} \\
\text{sem: pad} \\
\text{mdd:} \\
\text{mdr:} \\
\text{nc:} \\
\text{pc: big} \\
\text{prn: 23}
\end{bmatrix}
$$

$$
\begin{bmatrix}
\text{sur: lucy} \\
\text{noun: [person x]} \\
\text{cat: snp} \\
\text{sem: nm f} \\
\text{fnc: be} \\
\text{mdr:} \\
\text{nc:} \\
\text{pc:} \\
\text{prn: 24}
\end{bmatrix}
\begin{bmatrix}
\text{sur:} \\
\text{verb: be} \\
\text{cat: \#n' \#be' decl} \\
\text{sem: pres} \\
\text{arg: [person x]} \\
\text{mdr: happy} \\
\text{nc:} \\
\text{pc:} \\
\text{prn: 24}
\end{bmatrix}
\begin{bmatrix}
\text{sur:} \\
\text{adj: happy} \\
\text{cat: adn} \\
\text{sem: pad} \\
\text{mdd: be} \\
\text{mdr:} \\
\text{nc:} \\
\text{pc:} \\
\text{prn: 24}
\end{bmatrix} \dots
$$

The extrapropositional coordination is coded by the next conjunct feature [nc: (be 24)] of the predicate *find*.

15.9 Memory-Based Concatenation in Nonlanguage Recognition

The elementary features provided by the interface component of a cognitive agent may be combined in many different ways, in different modalities, and at different levels of complexity, creating a huge search space. In natural language communication this problem is solved efficiently by (i) the time-linear order of the *surface input* to the hear mode and (ii) the time-linear navigation along the semantic relations in the *content input* to the speak mode. They result in the linear degree of computational complexity in natural language communication, which effectively provides for real time processing in LAG/DBS (TCS'92).

In contrast, the cognitive processing of raw data input and output without an explicit specification of the processing order, for example in vision, requires alternative means of reducing the search space. In humans, they are of a cultural nature, namely the agents' knowledge of their current environment, for example, being in the kitchen, in the bath room, in a lecture hall, on the street, in a car, in a garden, in a laboratory, watching a base ball game, a Western, a SciFi movie, etc., which (i) are continuously acquired by the members of a society and (ii) effectively constrain the set of associated visual concepts and their interconnections. For example, if the agent is in the well-kept garage of an electric vehicle, there is normally no need to prepare for recognizing a can of motor oil.

The same restrictions are used in natural language communication, though to a much lesser extent. For example, the word match may be used in the context of lighting a fire or combining pieces of garment, bank for a financial institution, the territory alongside a river, or a place to sit. Compared to vision without domain restrictions, the number of language-dependent lexical ambiguities is tiny, but they are disambiguated in the same way, namely by the agents' awareness of the context of use in the culture.[11]

In short, the number of distinctions within a cultural context of use depends on the agent's gender, age, origin, education, and personal interest. Also, what looks the same for a layman may be obviously different, even critical, for the expert. Just as there are natural experts and natural laymen in a certain field of expertise, there may be artificial counterparts which all have the same visual equipment, but vary vastly in their cultural and scientific knowledge, and in the skill of their actions.

While finer and finer cultural distinctions in a domain and a semantic field are an important ingredient of better and better visual recognition, they cannot suffice alone. At the bottom level, vision must be grounded in science, for language and nonlanguage concepts alike. This raises the question of how to get from the most elementary features like the length and orientation of straight lines (Hubel and Wiesel 1962), the size and color of circles, ovals, and rectangles, etc., to cups, pails, and watering cans, for example.

In cognitive psychology, this question has been addressed by Biederman's (1987) *Recognition-by-Components* (RBC).[12]It is summarized by Kirkpatrick (2001) as follows:

> The major contribution of RBC is the proposal that the visual system extracts geons (or geometric ions) and uses them to identify objects. Geons are simple volumes such as cubes, spheres, cylinders, and wedges. RBC proposes that representations of objects are stored in the brain as structural descriptions. A structural description contains a specification of the object's geons and their interrelations (e.g. the cube is above the cylinder).
>
> ...
>
> The RBC view of object recognition is analogous to speech perception. A small set of phonemes are combined using organizational rules to produce millions of different words. In RBC, the geons serve as phonemes and the spatial interrelations serve as organizational rules. Biederman 1987 estimated that as few as 36 geons could produce millions of unique objects.

Consider the following examples of geons:

---

[11] In transgressive art, this is made aware by Meret Oppenheim's *Fur Cup* (1936) and in pop art by Claes Oldenburg's oversized *Plantoir* (2001).

15.9.1 A SMALL SET OF GEONS

The geons may be assembled into concepts for complex objects such as the following:

15.9.2 COMBINING THE GEONS INTO MORE COMPLEX OBJECTS

As complex objects, they raise the following question for DBS: How should the combinations of geons into concepts for complex objects be stored? For example, should the handle and the cylinder of a pail be represented adjacent, as in the graphical representation 15.9.2, or should we specify their connection in a more abstract manner, thus opening the way to use geons like proplets as the basic key for storage and retrieval in a DBS on-board database?

This question bears on an important task in visual recognition, namely *pattern completion* (Barsalou 1999). If visual recognition is incremental, such that we see some part first and then rapidly reconstruct the rest of the object by looking for known connections stored in memory, how can we get the database to provide relevant visual content fast and succinct to quickly narrow down the search?

It turns out that the data structure and the associated retrieval algorithm of DBS language cognition provide a highly efficient procedure of pattern completion in nonlanguage cognition. As an example, consider recognition of a pail based on a DBS A-memory containing the following geons:

15.9.3 PATTERN COMPLETION DURING RBC RECOGNITION

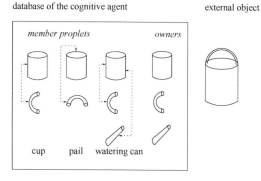

The isolated geons are the owners and the connected geons the members. The con-

---

[12] Thanks to Professor Brian MacWhinney for helpful suggestions and a three month visit at the CMU Psychology Department in 1989.

nections between the geons of each complex concept are indicated by dotted arrows. The complex objects specified in each column of connected geons are provided with names, here cup, pail, and watering can.

Given that the external object is a pail, the agent might either first recognize the handle or the cylinder – depending on the conditions of lighting or the orientation of the object. If the cylinder is recognized first, the agent's database will indicate that cylinders are known to be connected in certain ways to handles and/or spouts. This information is used to actively analyze the cylinder's relations to the rest of the external object (pattern completion), checking for the presence or absence of the items suggested by the data base.

Similarly, if the handle is recognized first, the agent's database indicates that handles are known to be connected in certain ways to cylinders to form cups, pails, or watering cans. If the handle-cylinder connection is recognized first, there are two possibilities: pail or watering can. In our example, the system determines that there is no spout and recognizes the object as a pail.

For specifying the connections between geons more precisely let us replace the intuitive graphical model illustrated in 15.9.3 with the DBS memory format based on variants of proplets. The following example expresses the same content, but represents geons by names, and codes the connections between geons by means of features (attribute value pairs).

15.9.4 STORING COMPLEX OBJECTS AS GEON PROPLETS

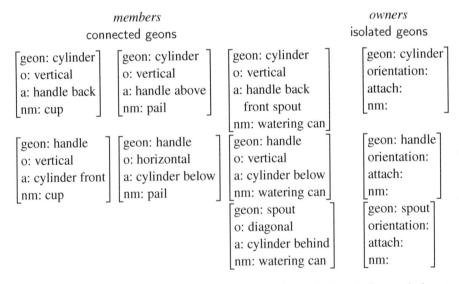

The attribute o stands for orientation, here *horizontal, vertical,* and *diagonal,* the attribute a for attachment, here *above, below, back,* and *front,* and the attribute nm for name, here *cup, pail,* and *watering can.*

In this small example with the owner geons cylinder, handle, and spout, the recognition algorithm works as follows. If the owner, for example cylinder, is matched by the raw data (data-driven activation), the algorithm checks its token line and finds three *cylinder* proplets with different nm values, namely cup, pail, and watering can, different a values, namely handle back, handle above, and handle above, spout front, and different o values, namely vertical, horizontal, and diagonal.

The *cup* proplet has the attachment value handle back. If the interface component analysis of the raw data confirms that the handle attaches to the back of the cylinder geon and there is no spout, recognition is complete. Otherwise the attachment value handle above of the *pail* proplet is checked. If it is found in the raw data to attach at the top of the cylinder geon, recognition is complete.

The *watering can* hypothesis is the worst case, because it is alphabetically last (assuming an alphabetical rather than frequency ordering of attachment attempts) and requires two matching checks. If none of the three hypotheses is confirmed, the recognition attempts fail and an additional complex concept is constructed.

In more general applications, efficiency may be improved by using token lines with a frequency-based order in combination with domain and semantic field restrictions. Also, raw data need not always be thoroughly recognized. For example, when looking for the watering can in the context of a cluttered garden shed much of the raw data may be left unanalyzed or be analyzed crudely. By recording the raw data without analysis, however, there may be recognition from memory at a later time (such as a detective's reasoning on his sofa in front of a blazing fire with a bottle of single malt).

15.10   Conclusion

The database schema, data structure, and algorithm of a relational database, e.g., an RDBMS, is *record-based*, while those of a DBS database are *content-based* and serve as the on-board memory of an artificial agent. The input-output conditions of a relational database are for storing and retrieving data, while those of a DBS database are for automatic (i) nonlanguage recognition and action, (ii) transfer of content from speaker to hearer in natural language communication, and (iii) reasoning

To show convergence, the chapter presents two applications which go beyond the transfer of content between agents by means of language-dependent raw surface data. One is the psychological phenomenon of association as a coactivation of resonating content in the on-board database, activated by content at the agent's current now front. The other is concatenation without an externally-given processing order in nonlanguage visual perception, using Biederman's (1987) 'Recognition-by-Components.

---

[12] See CC 13 for more on the coactivation of content by autonomous navigation.

16. David Hume's 'Causation' in Database Semantics

In physics, causation is an agent-external phenomenon, for example gravity causing the movement of the planets (Newton 1687). According to Hume's (1739) philosophical reception (Slavov 2013) of Newton's law of gravitation, the cause must precede the effect (called *temporal priority*), and cause and effect must be spatiotemporally conjoined (called *contiguity*).

In agent-internal cognition, there is an analogous phenomenon. For example, (x) Mary turned off the light and fell asleep satisfies temporal priority and contiguity, while (y) Mary fell asleep and turned off the light violates temporal priority. Content tokens matching the content types (x) and (y) are called *accommodating contents*, and content tokens satisfying content type (x) but not (y) are called *functionally accommodating* in DBS.

Functional accommodation is a generalization of Hume's cause and effect. For example, John put on his socks and shoes is an instance of functional accommodation though not of causation, while John put on his shoes and socks is a case of neither. This chapter presents the technical details of reconstructing functional accommodation in DBS.

16.1 Asymmetry in Natural Coordination

In propositional calculus, conjunction is symmetric, i.e. $p \wedge q = q \wedge p$, and implication is asymmetric, i.e. $p \rightarrow q \neq q \rightarrow p$. In natural cognition, however, coordination may be asymmetric, either because of (a) different functional accommodations or (b) their presence vs. absence. The following contents of clausal coordination represent alternative orders by means of English language surfaces:

16.1.1 IMPLICATIONS IN NATURAL COORDINATION CONTENTS

1. (x) John opened the window and threw out the cat.[1]
   (y) John threw out the cat and opened the window.
2. (x) John got ill and Mary made soup.
   (y) Mary made soup and John got ill.

---

[1] Prof. G. Lakoff, lecture at the Linguistic Summer School UC Santa Cruz, 1971 or 1972.

© The Author(s), under exclusive license to Springer Nature Switzerland AG 2023
R. Hausser, *Ontology of Communication*, https://doi.org/10.1007/978-3-031-22739-4_16

3. (x) Mary turned off the light and fell asleep.
   (y) Mary fell asleep and turned off the light.

4. (x) John put on the water and boiled the potatoes.
   (y) John boiled the potatoes and put on the water.

5. (x) Suzy opened the fridge and got a beer.
   (y) Suzy got a beer and opened the fridge.

6. (x) Suzy got a beer and closed the fridge.
   (y) Suzy closed the fridge and got a beer.

The alternative conjunction orders in examples (1) and (2) support different functional accommodations. More specifically, order (x) in (1) implies that John opened the window for the purpose of throwing the cat through it while order (y) may be instantiated by a more conventional exit of the cat having caused bad air. The order (x) in (2) may be motivated as Mary's intention to comfort John, while on order (y) John's illness may have been caused by Mary's soup. The respective orders in the remaining examples (3–6), in contrast, support functional accommodation in the variants (x), while there is no functional accommodation for the variants (y).

16.2 Cause and Effect

Except for 2.(y), the examples of functional accommodation in 16.1.1 are not instances of causation, yet they satisfy Hume's (1739) definition in terms of contiguity and temporal priority:

16.2.1 HUME'S DEFINITION OF CAUSATION:

> X causes Y if and only if the two events are spatiotemporally conjoined (contiguity), and X precedes Y (temporal priority),

For example, the contents (a) Suzy opened the fridge and (b) got a beer (16.1.1, 5.(x)) do not express a causal relation, but are spatiotemporally conjoined by the assumption (i) that the beer is located in the fridge (which makes sense in cultures with refrigerators and bottled beer) and (ii) the time intervals are adjacent. Spatio-temporal priority is fulfilled if the purpose of opening the fridge is getting the beer.

In summary, the examples in 16.1.1 show that contiguity and temporal priority may be fulfilled not only by the agents' cognition-external reality (e.g., gravity), but also by accommodating contents which consist of two clauses conjoined in a certain order. The examples show also that accommodating contents are not limited to causation, but include all kinds of regular interactions, such as purpose and natural order, for example, putting on the socks before the shoes, slowing down before getting off the bike, digging the foundation before putting on the roof, etc.

16.3 Necessary, Unnecessary, Sufficient, and Insufficient Causes

Hume requires *constant* contiguity for cause and effect, i.e. Y must *always* follow from X, while for functional accommodation in DBS sporadic consequents are sufficient. For Hume's causation, constant contiguity is widely accepted as a necessary condition, but whether it is also sufficient is controversial.

For a more differentiated account of complex causes, J.L. Mackie (1965) distinguishes (i) necessary, (ii) unnecessary, (iii) sufficient, and (iv) insufficient causes, called the INUS condition by Mackie. For example, a short circuit causing a house on fire is a US constellation: the short circuit is Unecessary (what Aristotle calls *accidental*[2]) because there are other possible causes, such as arson or lightning. The short circuit is Sufficient because it effectively caused the house to burn.

16.4 Hume's Copy Principle

Underlying Hume's definition 16.2.1 is his *copy principle*[3], according to which the efficacy of elementary ideas comes from impressions copied into the mind, while the efficacy of the combination of elementary ideas into complex ideas is provided by the mind alone. For example, the elementary impressions *golden* and *mountain* and their corresponding ideas have counterparts in the real world, while the complex idea *golden mountain* does not, and similarly for *Pegasus* and *unicorn*.

In terms of agent-based data-driven DBS, this would mean that the recognition and action of elementary concepts is provided by the agent's interface component, but their combination into complex content is entirely cognition-internal. In fact, however, no such distinction is made in DBS.

As shown by such phenomena as visual illusion and mishearing (recognition), as well as mishandling and losing one's way (action), the type-token matching between elementary contents (concepts) and raw data (e.g., sound or light waves) is no less cognition-based than their combination by the semantic relations of structure, for example in inferencing. In other words, DBS agrees with Hume in that the combination of elementary ideas into complex ideas is provided by the mind, i.e. the agent's cognition. It is just that elementary recognition and action in DBS are provided by the mind as well. Also, the functional accommodation of DBS is not limited to causality, but generalizes to a multitude of other systematic relations.

16.5 Reconstruction of Elementary Recognition and Action

Hume described the 'mind' in terms of 'impressions' and 'ideas.' Impressions are divided into 'sensations' and 'reflections.' In the computational cognition of DBS, in

---

[2] Whether the opposite of Aristotle's accidental is *essential* or *necessary* is hotly debated in philosophy (Matthews 1990). We follow Quine 1966 by treating 'essential' and 'necessary' as equal.

[3] Unlike his contemporaries who still published in Latin, Hume published in English.

contrast, the most basic distinction is between *recognition* and *action*. These notions are absent in Hume's ontology (Johansson 2012).

The systematic reconstruction of recognition and action in DBS is based on the distinction between 'types' and 'tokens' (Peirce 1906, CP Vol.4, p. 375), which goes back to Aristotle's distinction between the necessary and the accidental.

16.5.1 TYPE-TOKEN MATCHING FOR RECOGNITION AND ACTION IN DBS

- recognition: a type matching raw data results in a token.

- action: adapting a type into a token for a purpose results in raw data.

In other words, while for Hume the operations of the mind are founded on simple impressions which are received passively,[4] the recognition and action of basic concepts by the computational cognition of agent-based data-driven DBS is *proactive*.

As an example consider a DBS agent's recognizing and producing a square:

16.5.2 RECOGNITION OF A square

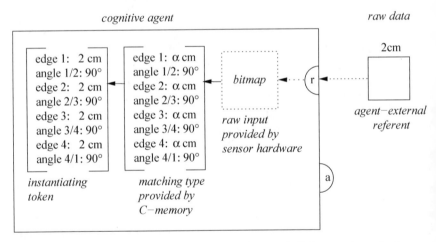

The edge length of the type is a variable which matches an infinite number of tokens with different edge lengths. The raw data are supplied by a sensor, here for vision, as input to the agent's interface component.

In action, a type is adapted to a token for the purpose at hand and realized by the agent's actuators as raw data:

---

[4] Though with "most force and violence" (Hume 1739).

16.5.3 ACTION OF REALIZING square

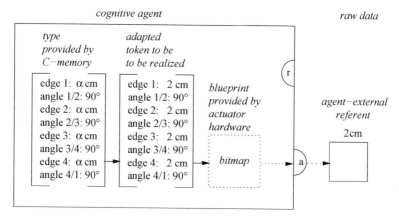

The token is used as a blueprint for action (e.g., drawing a square). The recognition and production of square may be extended to all two-dimensional geometric shapes (Hausser 2021b, 10.3.5)

Next consider the recognition of a color, here blue:

16.5.4 RECOGNITION OF blue

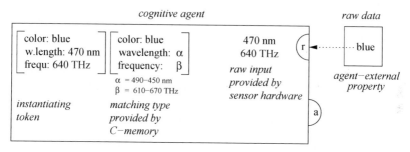

An example of the corresponding action is turning on the color blue, as in a cuttlefish (metasepia pfefferi) using its chromatophores:

16.5.5 ACTION OF REALIZING blue

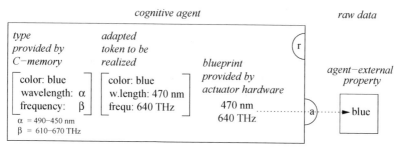

The concept type matches different shades of blue, whereby the variables α and β are instantiated as constants in the resulting token. Recognition and production of blue is a general mechanism which may be applied to all colors (Hausser 2021b, 10.3.1). It may be expanded to infrared and ultraviolet, and to varying intensity.

16.6 Computational Reconstruction of Complex Content

In DBS, concepts are embedded as core values into nonrecursive feature structures with ordered attributes, called *proplets*:

16.6.1 LEXICAL PROPLETS OF blue AND square

$$
\begin{bmatrix}
\text{sur:} \\
\text{adj: } \textbf{blue} \\
\text{cat: adn} \\
\text{sem: pad} \\
\text{mdd:} \\
\text{mdr:} \\
\text{nc:} \\
\text{pc:} \\
\text{prn: K}
\end{bmatrix}
\qquad\qquad
\begin{bmatrix}
\text{sur:} \\
\text{noun: } \textbf{square} \\
\text{cat: snp} \\
\text{sem: indef sg} \\
\text{fnc:} \\
\text{mdr:} \\
\text{nc:} \\
\text{pc:} \\
\text{prn: K}
\end{bmatrix}
$$

Proplets are the computational data structure of DBS. Their second attribute is the core attribute, here adj and noun, and contains the core value. Their fifth attribute is the continuation attribute, here mdd for 'modified' and fnc for 'functor', intended for the continuation value. The cat and sem slots provide the syntactic and the semantic properties of the concept.

Proplets are connected into complex content by cross-copying between their core and their continuation attributes, shown in **bold face** in the following example:

16.6.2 'LUCY FOUND A BIG BLUE SQUARE.' AS NONLANGUAGE CONTENT

$$
\begin{bmatrix}
\text{sur: lucy} \\
\text{noun: } \textbf{[person x]} \\
\text{cat: snp} \\
\text{sem: nm f} \\
\text{fnc: } \textbf{find} \\
\text{mdr:} \\
\text{nc:} \\
\text{pc:} \\
\text{prn: 23}
\end{bmatrix}
\begin{bmatrix}
\text{sur:} \\
\text{verb: } \textbf{find} \\
\text{cat: } \#n' \#a' \text{ decl} \\
\text{sem: ind past} \\
\text{arg: } \textbf{[person x] square} \\
\text{mdr:} \\
\text{nc:} \\
\text{pc:} \\
\text{prn: 23}
\end{bmatrix}
\begin{bmatrix}
\text{sur:} \\
\text{adj: } \textbf{big} \\
\text{cat: adn} \\
\text{sem: pad} \\
\text{mdd: } \textbf{square} \\
\text{mdr:} \\
\text{nc: } \textbf{blue} \\
\text{pc:} \\
\text{prn: 23}
\end{bmatrix}
\begin{bmatrix}
\text{sur:} \\
\text{adj: } \textbf{blue} \\
\text{cat: adn} \\
\text{sem: pad} \\
\text{mdd:} \\
\text{mdr:} \\
\text{nc:} \\
\text{pc: } \textbf{big} \\
\text{prn: 23}
\end{bmatrix}
\begin{bmatrix}
\text{sur:} \\
\text{noun: } \textbf{square} \\
\text{cat: snp} \\
\text{sem: indef sg} \\
\text{fnc: } \textbf{find} \\
\text{mdr: } \textbf{big} \\
\text{nc:} \\
\text{pc:} \\
\text{prn: 23}
\end{bmatrix}
$$

The example is a nonlanguage content because the sur slots are either empty or a name marker. The explicit definitions of the values blue and square is shown in 16.5, and similar definitions are assumed for the other core values. The semantic relations are classical subject/predicate, object\predicate, modifier|modified, and conjunct−conjunct. They are established by the cross-copying (connective ×) or the absorption (connective ∪) of values (TExer).

The semantic relations in 16.6.2, i.e. person(x)/find, square\find, big|square, and big−blue, may be shown graphically as follows:

16.6.3 SEMANTIC RELATIONS UNDERLYING SPEAK MODE DERIVATION

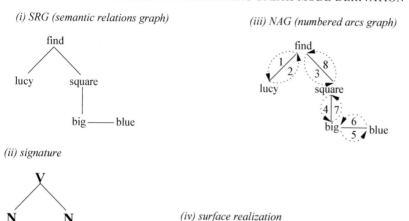

(i) SRG (semantic relations graph)

(iii) NAG (numbered arcs graph)

(ii) signature

(iv) surface realization

1	2	3	4	5	6	7	8
Lucy	found	a	big	blue		square	.
V/N	N/V	V\N	N|A	A–A	A–A	A|N	N|V

Recognition takes a linear sequence of connected proplets, e.g., 16.6.2, as input and produces an equivalent semantic hierarchy, e.g., 16.6.3, as output by interpreting the semantic relations encoded by cross-copying. Action takes a semantic hierarchy, e.g., 16.6.3, as input and produces a linear sequence, e.g., 16.6.2, as output by navigating along the semantic relations as shown in the NAG. The proplets of a content are connected by the semantic relations of structure and their prn value, but order-free for storage in and retrieval from the agent's content-addressable on-board database.

16.7 From Individual Contents to a Content Class

The equivalent syntactic-semantic structure of 16.6.2 and 16.6.3 is the same for an unlimited number of contents which differ solely in their core and continuation values, as shown by the following example:

16.7.1 CONTENT IN THE SAME CLASS AS 16.6.2 AND 16.6.3

Peter ate a sweet little chocolate.

The class containing Lucy found a big blue square and Peter ate a sweet little chocolate may be characterized abstractly by the same schema, derived from 16.6.1 by simultaneous substitution:

16.7.2 CONTENT SCHEMA AS A SET OF PROPLET PATTERNS

sur:	sur:	sur:	sur:	sur:
noun: α	verb: β	adj: γ	adj: δ	noun: ε
cat: snp	cat: #n′ #a′ decl	cat: adn	cat: adn	cat: snp
sem: nm f	sem: ind past	sem: pad	sem: pad	sem: indef sg
fnc: β	arg: $\alpha \varepsilon$	mdd: ε	mdd:	fnc: β
mdr:	mdr:	mdr:	mdr:	mdr: γ
nc:	nc:	nc: δ	nc:	nc:
pc:	pc:	pc:	pc: γ	pc:
prn: K	prn: K	prn: K	prn: K	prn: K

This syntactic-semantic schema of a content resulted from 16.6.2 by simultaneously substituting the values in bold face with Greek letters representing variables.

The analogous method is also applied to generalize the graphical format of 16.6.3 from an individual instance to the class:[5]

16.7.3 CONTENT SCHEMA AS SEMANTIC RELATIONS GRAPHS

(i) SRG (semantic relations graph)   *(iii) NAG (numbered arcs graph)*

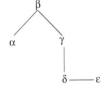

These graphical representations of semantic relations characterize the abstract class which matches the contents of Lucy found a big blue square. Peter ate a sweet little chocolate., and an open number of similar constructions.

16.8 Four Different Kinds of Content

DBS applies the type-token distinction not only to concepts (16.5) but also to content. In combination with the nonlanguage-language distinction there are four kinds of content in DBS, called [−surface −STAR], [−surface +STAR], [+surface −STAR], and [+surface +STAR], illustrated as follows:

16.8.1 NONLANGUAGE CONTENT TYPE: [−surface, −STAR]

sur:	sur:	sur:
noun: dog	verb: find	noun: bone
cat: snp	cat: #n′ #a′ decl	cat: snp
sem: def sg	sem: past ind	sem: indef sg
fnc: find	arg: dog bone	fnc: find
mdr:	mdr:	mdr:
nc:	nc:	nc:
pc:	pc:	pc:
prn: K	prn: K	prn: K

This proposition is a type because there is no STAR and the prn value is a variable, here K. It is a nonlanguage content because the sur slots are empty.

The next example is a corresponding nonlanguage token:

16.8.2 NONLANGUAGE CONTENT TOKEN: [−surface, +STAR]

$$
\begin{bmatrix} \text{sur:} \\ \text{noun: dog} \\ \text{cat: snp} \\ \text{sem: def sg} \\ \text{fnc: find} \\ \text{mdr:} \\ \text{nc:} \\ \text{pc:} \\ \text{prn: 12} \end{bmatrix}
\begin{bmatrix} \text{sur:} \\ \text{verb: find} \\ \text{cat: } \#n' \ \#a' \ \text{decl} \\ \text{sem: past ind} \\ \text{arg: dog bone} \\ \text{mdr:} \\ \text{nc:} \\ \text{pc:} \\ \text{prn: 12} \end{bmatrix}
\begin{bmatrix} \text{sur:} \\ \text{noun: bone} \\ \text{cat: snp} \\ \text{sem: indef sg} \\ \text{fnc: find} \\ \text{mdr:} \\ \text{nc:} \\ \text{pc:} \\ \text{prn: 12} \end{bmatrix}
\begin{bmatrix} \text{S: yard} \\ \text{T: friday} \\ \text{A: sylvester} \\ \text{R:} \\ \text{3rd:} \\ \text{prn: 12} \end{bmatrix}
$$

The three content proplets and the STAR proplet are connected by a common prn constant, here 12. According to the STAR, the token resulted as an observation by the agent Sylvester on friday in the yard.

The following language content type corresponding to 16.8.1 illustrates the independence of language-dependent sur values, here German, from the relatively language-independent placeholders for concepts (represented by English base forms for convenience):

16.8.3 Language content type: [+surface, −STAR]

$$
\begin{bmatrix} \text{sur: der\_Hund} \\ \text{noun: dog} \\ \text{cat: snp} \\ \text{sem: def sg} \\ \text{fnc: find} \\ \text{mdr:} \\ \text{nc:} \\ \text{pc:} \\ \text{prn: K} \end{bmatrix}
\begin{bmatrix} \text{sur: fand} \\ \text{verb: find} \\ \text{cat: } \#n' \ \#a' \ \text{decl} \\ \text{sem: past ind} \\ \text{arg: dog bone} \\ \text{mdr:} \\ \text{nc:} \\ \text{pc:} \\ \text{prn: K} \end{bmatrix}
\begin{bmatrix} \text{sur: einen\_Knochen} \\ \text{noun: bone} \\ \text{cat: snp} \\ \text{sem: indef sg} \\ \text{fnc: find} \\ \text{mdr:} \\ \text{nc:} \\ \text{pc:} \\ \text{prn: K} \end{bmatrix}
$$

A language content type is also called a literal meaning[1]. It is an abstraction in that an actual DBS hear mode derivation results in a content token. However, a content type may always be obtained from a content token by removing the STAR and replacing the prn constants with appropriate variables.

The fourth kind of content is a language token which matches the type, here 16.8.3, called an utterance meaning[2]. The example is produced by the speaker Sylvester in German towards the intended hearer Tweety and corresponds to the nonlanguage content token 16.8.2 except for the R value:

---

[5] Formats *(ii)* and *(iv)* omitted.

16.8.4 LANGUAGE CONTENT TOKEN: [+surface, +STAR]

⌈sur: der_Hund⌉	⌈sur: fand	⌉	⌈sur: einen_Knochen⌉	⌈S: yard	⌉
noun: dog	verb: find		noun: bone	T: friday	
cat: snp	cat: #n′ #a′ decl		cat: snp	A: sylvester	
sem: def sg	sem: past ind		sem: indef sg	R: tweety	
fnc: find	arg: dog bone		fnc: find	3rd:	
mdr:	mdr:		mdr:	⌊prn: 12	⌋
nc:	nc:		nc:		
pc:	pc:		pc:		
⌊prn: 12 ⌋	⌊prn: 12	⌋	⌊prn: 12 ⌋		

According to the STAR, the transfer of content occurred in the yard on friday from Sylvester to Tweety. The content types 16.8.1 and 16.8.3 match not only the tokens 16.8.2 and 16.8.4, but an open number of corresponding tokens with different prn values.

An utterance meaning$_2$ exists in the cognition of the speaker, and – if transfer is successful – of the hearer. The raw data serving as the vehicle of transfer in communication, in contrast, have absolutely no meaning or grammatical properties whatsoever at all (no reification in DBS), but may be measured by natural science.

16.9 Accommodating Scenarios in DBS

The DBS notion of a *complex content* as a set (order-free) of proplets connected by the semantic relations of structure is essential for the computational implementation of accommodating scenarios in general and functional accommodation in particular. As an example consider 5(x) in 16.1.1 as a DBS content:

16.9.1 CONTENT TOKEN OF Suzy opened the fridge and got a beer.

⌈sur: suzy ⌉	⌈sur:	⌉	⌈sur:	⌉	⌈sur:	⌉	⌈sur:	⌉	⌈S: kitchen⌉
noun: [p. x]	verb: open		noun: fridge		verb: get		noun: beer		T: 6pm
cat: snp	cat: #s3′ #a′ decl		cat: snp		cat: #s3′ #a′ decl		cat: snp		A: Peter
sem: sg m	sem: *and* ind past		sem: def sg		sem: ind past		sem: indef sg		R: Lizzy
fnc: open	arg: [p. x] fridge		fnc: open		arg: [p. x] beer		fnc: get		3rd:
mdr:	mdr:		mdr:		mdr:		mdr:		⌊prn: 23 ⌋
nc:	nc: (get 24)[6]		nc:		nc:		nc:		
pc:	pc:		pc:		pc:		pc:		
⌊prn: 23 ⌋	⌊prn: 23 ⌋		⌊prn: 23 ⌋		⌊prn: 24 ⌋		⌊prn: 24 ⌋		

This content is a token because of the explicit STAR. Temporal priority is encoded by the consecutive prn values 23 and 24. Contiguity is supported intuitively by the content of the two clausal conjuncts (coactivation, Hausser 2021c, 15.6–15.8).

The abstract syntactic-semantic structure of this content is shared by all the other (x)-variants in 16.1.1 and may be characterized as the following schema:

---

[6] In extrapropositional coordination, the forward direction is implemented routinely in DBS, while the backward direction is handled by an inference which applies only when needed and provides the necessary conjunctions such as before that or earlier (Hausser 2021d, 5.5).

16.9.2 CLAUSAL COORDINATION WITH FUNCTIONAL ACCOMMODATION

$$
\begin{bmatrix} \text{noun: } \alpha \\ \text{fnc: } \beta \\ \text{prn: K} \end{bmatrix}
\begin{bmatrix} \text{verb: } \beta \\ \text{sem: } and \\ \text{arg: } \alpha\ \gamma \\ \text{nc: } \delta\ \text{K+1} \\ \text{prn: K} \end{bmatrix}
\begin{bmatrix} \text{noun: } \gamma \\ \text{fnc: } \beta \\ \text{prn: K} \end{bmatrix}
\begin{bmatrix} \text{verb: } \delta \\ \text{arg: } \alpha\ \varepsilon \\ \text{prn: K+1} \end{bmatrix}
\begin{bmatrix} \text{noun: } \varepsilon \\ \text{fnc: } \delta \\ \text{prn: K+1} \end{bmatrix}
\begin{bmatrix} \text{S: q} \\ \text{T: r} \\ \text{A: s} \\ \text{R: t} \\ \text{3rd:} \\ \text{prn: K} \end{bmatrix}
$$

where β precedes δ, and β and δ are contiguous.

The schema 16.9.2 is derived from the content 16.9.1 by simultaneous substitution of the core and continuation values with variables represented by Greek letters. 16.9.2 matches all contents with the same syntactic structure as 16.9.1, for example, Mary turned_off the light and fell asleep. The syntactic-semantic structure of clausal coordination is specified by the abstract patterns of the schema 16.9.2. Functional accommodation (as a generalization of Hume's definition 16.2.1) is a matter of the prn values and the STAR.

16.10   Conclusion

Accommodating scenarios are based on the agents' cultural background and personal experiences. Stored in the agents' content-addressable on-board database (memory) and actived (Hausser 2021d, 5.2–5.5) by current nonlanguage and language content processing, accommodating scenarios are an important ingredient of 'making sense.' Speaker and hearer activating the same accommodating scenarios supports reciprocal understanding in natural language communication.

A special case of accommodating scenarios is *functional accommodation*. Syntactically, functional accommodation requires clausal coordination and a certain order of the clausal conjuncts. Semantically it requires the spatiotemporal contiguity of the conjuncts. This equals Hume's (1739) definition of causation.

However, while Hume's causation applies to the agent-external reality (e.g., gravity), functional accommodation applies to the agent-internal cognition of DBS, which is agent-based data-driven. Because the agent-external reality has necessarily[7] an agent-internal cognitive aspect in DBS, Hume's causation is subject to functional accommodation as well. In this sense, functional accommodation may be viewed as a generalization of Hume's causation.

Technically, DBS cognition is based on an operational analysis of concepts in terms of the agents' recognition and action, the computational data structure of proplets, computational pattern matching between types and tokens (in concepts, proplets, and contents), operations which use the cross-copying of values to establish the semantic relations of structure between the concepts embedded in proplets, and the time, space, and agent information coded in the STAR of clausal content tokens. It is shown that

---

[7] In as much as it is conscious.

Hume's account of causality in terms of agent-external temporal priority and contiguity may be generalized and reconstructed computationally in agent-based data-driven DBS without any need for auxiliary additions.

17. Concepts in Computational Cognition

The sign kinds of natural language are the *concepts*, the *indexicals,* and the *names*. Of these, only the concepts interact directly with the agent's cognition-external environment, whereas indexicals and names receive their interpretation indirectly from cognition-internal content.

The analysis of concepts has been based on the ten categories of Aristotle, the four categories of Kant, Wittgenstein's family resemblance, and the prototypes of cognitive psychology. Their computational implementation as the recognition and action of an artificial cognition is adequate if and only if they equal the natural counterpart.

In natural and technological concepts, the desired equivalence has solutions grounded in science.[1] The problem is the computational implementation of the cultural concepts in different belief systems and traditions. The technical details of content transfer from speaker to hearer by means of raw data (sound waves, formants, light waves, pixels) are explicated in Sect. 17.8.

17.1 Concept-Based Interpretation of Indexicals and Names

In agent-based data-driven DBS (AIJ'92), the reliance of indexicals and names on concepts is based on the type-token distinction from philosophy (Peirce 1906, CP Vol.4, p. 375) which goes back to Aristotle's distinction between the necessary and the accidental (Metaphysics). Consider the DBS analysis of a nonlanguage clausal content type with an indexical (first person pronoun pro1) and a name (Fido):

17.1.1 NONLANGUAGE CONTENT OF I saw Fido. AS TYPE

```
┌sur:       ┐ ┌sur:          ┐ ┌sur: fido┐
│noun: pro1 │ │verb: see     │ │noun: β  │
│cat: s1    │ │cat: #n #a decl│ │cat: snp │
│sem: sg    │ │sem: ind past │ │sem: sg  │
│fnc: see   │ │arg: pro1 β   │ │fnc: see │
│mdr:       │ │mdr:          │ │mdr:     │
│nc:        │ │nc:           │ │nc:      │
│pc:        │ │pc:           │ │pc:      │
└prn: K     ┘ └prn: K        ┘ └prn: K   ┘
```

This is a nonlanguage content because the first two sur slots are empty and the value of the third sur slot is a marker, here fido (needed for the speak mode of agent-based DBS). It is a type because it is not connected to a STAR (as provided by the agent's onboard orientation system), and the core value of the name proplet and the prn values are variables, here β and K.

---

[1] For the grounding of concepts in computer science see Barsalou et al. (2003) and Steels (2008).

© The Author(s), under exclusive license to Springer Nature Switzerland AG 2023
R. Hausser, *Ontology of Communication*, https://doi.org/10.1007/978-3-031-22739-4_17

The STAR of a language content specifies the value of Space (location of the speaker), Time (moment of utterance), Agent (speaker), and Recipient (hearer), plus 3rd (third person), and prn (proposition number). Based on a STAR and language-dependent sur values, the nonlanguage clausal content type 17.1.1 may be turned into the following token of a clausal language content:

17.1.2 LANGUAGE CONTENT OF I saw Fido. AS TOKEN

$$
\begin{bmatrix}
\text{sur: I} \\
\text{noun: } \mathbf{pro1} \\
\text{cat: snp} \\
\text{sem: sg} \\
\text{fnc: see} \\
\text{mdr:} \\
\text{nc:} \\
\text{pc:} \\
\text{prn: 3}
\end{bmatrix}
\begin{bmatrix}
\text{sur: saw} \\
\text{verb: see} \\
\text{cat: \#n \#a decl} \\
\text{sem: ind past} \\
\text{arg: } \mathbf{pro1} \text{ } [\mathbf{dog\ x}] \\
\text{mdr:} \\
\text{nc:} \\
\text{pc:} \\
\text{prn: 3}
\end{bmatrix}
\begin{bmatrix}
\text{sur: Fido} \\
\text{noun: } [\mathbf{dog\ x}] \\
\text{cat: sp2} \\
\text{sem: sg} \\
\text{fnc: see} \\
\text{mdr:} \\
\text{nc:} \\
\text{pc:} \\
\text{prn: 3}
\end{bmatrix}
\begin{bmatrix}
\text{S: backyard} \\
\text{T: Monday} \\
\text{A: Sylvester} \\
\text{R: Speedy} \\
\text{3rd:} \\
\text{prn: 3}
\end{bmatrix}
$$

This is a language content because the sur slots have surfaces as values, here English. It is a token because it is connected to an explicit STAR proplet by a shared prn value defined as a constant, here 3, and the name proplet has a named referent (CASM'17) as core value, here [dog x], instead of a variable. According to the STAR, the sentence was uttered in the space S (backyard) at the time T (Monday) by the agent A (speaker Sylvester) directed at the recipient R (hearer Speedy).

17.1.2 illustrates the dependence of indexicals on concepts by **pro1** pointing at the A value of the STAR. It shows the dependence of names on concepts by the named referent [dog x] as the core value of Fido, which serves as the grammatical object (second arg value of **see**). The Space and Time values of the STAR instantiate Aristotle's and Kant's category of *quantity* in DBS.

17.2 Concepts Grounded in Science

The type-token distinction applies not only to clausal, but also to phrasal and elementary content. In elementary concepts with a grounding in physics this is straightforward, such as the following concept of the color blue:

17.2.1 TYPE AND TOKEN OF THE COLOR CONCEPT blue

type

$$
\begin{bmatrix}
\text{place holder: blue} \\
\text{sensory modality: vision} \\
\text{semantic field: color} \\
\text{content kind: concept} \\
\text{wavelength: 450–495nm} \\
\text{frequency: 670–610 THz} \\
\text{samples: a, b, c, ...}
\end{bmatrix}
$$

token

$$
\begin{bmatrix}
\text{place holder: blue} \\
\text{sensory modality: vision} \\
\text{semantic field: color} \\
\text{content kind: concept} \\
\text{wavelength: 470nm} \\
\text{frequency: 637 THz}
\end{bmatrix}
$$

In the type, the color is specified by intervals for wavelength and frequency. In the token, the intervals are replaced by constants which lie within the intervals.

This method of defining the color blue may be generalized to all colors:

17.2.2 SIMILARITY AND DIFFERENCE BETWEEN COLOR CONCEPT TYPES

⎡place holder: red sensory modality: vision semantic field: color content kind: concept wavelength: 700-635 nm frequency: 430-480 THz samples: a, b, c, ...⎤	⎡place holder: green sensory modality: vision semantic field: color content kind: concept wavelength:495-570 nm frequency: 526-606 THz samples: a′, b′, c′, ...⎤	⎡place holder: blue sensory modality: vision semantic field: color content kind: concept wavelength: 490-450 nm frequency: 610-670 THz samples: a″, b″, c″, ...⎤

The three types differ in their wavelength and frequency intervals, and their place holder and samples values; they share the sensory modality, semantic field, and content kind values.

Another class of concepts grounded in science are the shapes of two-dimensional geometry, such as the concept type and token of square:

17.2.3 TYPE AND TOKEN OF THE CONCEPT square

type

⎡place holder: square
sensory modality: vision
semantic field: two-dim geom.
content kind: concept
shape: ⎡edge 1: α cm
angle 1/2: 90°
edge 2: α cm
angle 2/3: 90°
edge 3: α cm
angle 3/4: 90°
edge 4: α cm
angle 4/1: 90°⎤
samples: a, b, c,...⎤

token

⎡place holder: square
sensory modality: vision
semantic field: two-dim geom.
content kind: concept
shape: ⎡edge 1: 2 cm
angle 1/2: 90°
edge 2: 2 cm
angle 2/3: 90°
edge 3: 2 cm
angle 3/4: 90°
edge 4: 2 cm
angle 4/1: 90°⎤⎤

The edge value of the type is the variable α which matches an infinite number of square tokens with different edge lengths, here 2cm in the token.

Just as the definition of the concept *blue* may be generalized routinely to other colors (17.2.2), the definition of the concept *square* may be generalized to other shapes in two-dimensional geometry, such as *equilateral triangle,* and *rectangle*:

17.2.4 SIMILARITY AND DIFFERENCE BETWEEN CONCEPT SHAPE TYPES

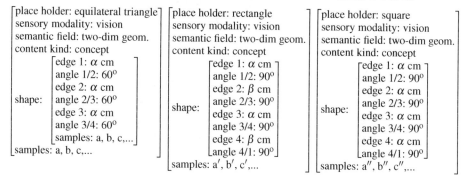

The operational implementation of color and two-dimensional geometric shape recognition and action is essential for building the computational cognition of a DBS robot (embodiment, MacWhinney 2008). For example, assuming eye-hand orientation, the robot could effectively pick blue squares from any sample of geometric shapes in different colors.

17.3 'Natural Categories' as Concepts

The concepts grounded in science, e.g. 17.2.2 and 17.2.4, are treated by Rosch (1973) as a subclass of the 'natural categories,' called *physiological categories*. The focus, however, is on categories like fruit, which are not 'physiological'. Based on psychological tests, Rosch shows empirically that the elements dominated by a higher category are not sets (unordered), but cognitively structured around a culture-dependent *prototype*.

For example, for most people in Western Europe the prototype dominated by fruit is apple, surrounded by plums, pines, and olives as less typical representatives, with a decrease in prototypicality from left to right (Rosch 1973: 130ff.). This prototype information of 'fruit' differs markedly from the biological definition:

17.3.1 BIOLOGY-BASED LEXICAL DEFINITION OF 'FRUIT'

> The fleshy or dry ripened ovary of a flowering plant, enclosing the seed or seeds. Thus, apricots, bananas, and grapes, as well as bean pods, corn grains, tomatoes, cucumbers, and (in their shells) acorns and almonds, are all technically fruits.
>
> Encyclopedia Britannica

The DBS definition of concepts as *nonrecursive* feature structures with *ordered* attributes (like proplets) is a simple and efficient computational format for combining (i) well-established lexical definitions, including those grounded in science ('physiological categories'), with (ii) prototypes and their (iii) satellites:

17.3.2 TYPE AND TOKEN OF THE CONCEPT 'FRUIT' IN DBS FORMAT

type

$$\begin{bmatrix} \text{placeholder: fruit} \\ \text{part of: flowering plant} \\ \text{prototype: apple} \\ \text{satellites: plums, pines, olives} \\ \text{use: edible} \\ \text{samples: a, b, c,...} \end{bmatrix}$$

token

$$\begin{bmatrix} \text{placeholder: fruit} \\ \text{part of: flowering plant} \\ \text{prototype: apple} \\ \text{instantiation: plum} \\ \text{use: edible} \\ \text{samples: b} \end{bmatrix}$$

This combination of (i) the lexical definition 17.3.1 via the place holder fruit, (ii) the cognitive prototype apple and (iii) its satellites supports different kinds of reasoning.

17.4 Technical Concepts as a Subclass of 'Natural Categories'

Another subclass of elementary concepts besides the physics-based (17.2.2, 17.2.4) and the biology-based (17.3.1) are the technological-based, for example, the concept of airplane. Lexically, the concept airplane has been defined follows:

17.4.1 LEXICAL DEFINITION OF 'AIRPLANE'

> Also called aeroplane or plane, any of a class of fixed-wing aircraft that is heavier than air, propelled by a screw propeller or a high-velocity jet, and supported by the dynamic reaction of the air against its wings.
>
> Encyclopedia Britannica

In analogy to the transition from 17.3.1 to 17.3.2, this definition may be integrated into the following nonrecursive feature structures with ordered attributes:

17.4.2 DBS CONCEPT TYPE AND TOKEN OF 'AIRPLANE'

type
$$\begin{bmatrix} \text{placeholder: airplane} \\ \text{part of: top node}^2 \\ \text{prototype: Boeing 373 MAX} \\ \text{satellites: Airbus A320,} \\ \quad \text{Cessna 172, Diamond DA40 NG, ...} \\ \text{use: transport} \\ \text{samples: a, b, c,...} \end{bmatrix}$$

token
$$\begin{bmatrix} \text{placeholder: airplane} \\ \text{part of: top node} \\ \text{prototype: Boeing 373 MAX} \\ \text{instantiation: Cessna 172} \\ \text{use: transport} \\ \text{sample: b} \end{bmatrix}$$

A cognitive prototype and its satellites are culture dependent and a statistical foundation alone is likely to be unsuitable for a well-functioning computational cognition. It is therefore advisable to equip a talking DBS robot with both, prototypes as well as well-established lexical definitions, which is easy enough (17.3.2, 17.4.2).

17.5 Grammatical Categories

The *categories* of philosophy and cognitive psychology are called *concepts* in linguistics, which uses the term category for the *grammatical* categories. The DBS data structure of proplets specifies the basic categories with the core attributes, i.e., noun, verb, or adj. These are differentiated further by the values of the cat and sem attributes. The combination of the core attribute and the cat and sem features in a proplet is called the *category complex* in DBS. In the following content (set of concatenated proplets), the category complexes are shown in *italics*:

---

[2] 'Vehicle' in the context of airplanes seems to be reserved for 'unmanned aerial vehicle' (UAV).

17.5.1 CATEGORY COMPLEXES IN Lucy found a big blue square.

sur: Lucy	sur: found	sur: big	sur: blue	sur: square
noun: [person x]	*verb:* find	*adj:* big	*adj:* blue	*noun:* square
cat: snp	*cat: #n' #a' decl*	*cat: adn*	*cat: adn*	*cat: snp*
sem: nm f	*sem: ind past*	*sem: pad*	*sem: pad*	*sem: indef sg*
fnc: find	arg: [person x] square	mdd: square	mdd:	fnc: find
mdr:	mdr:	mdr:	mdr:	mdr: big
nc:	nc:	nc: blue	nc:	nc:
pc:	pc:	pc:	pc: big	pc:
prn: K	prn: K	prn: K	prn: K	prn: K

The core values *blue* and *square* are defined in 17.2.1 and 17.2.3. For the other core
values in 17.5.1, i.e., [person x], *find*, and *big*, explicit definitions must be assumed.

17.6 Hear Mode: Concatenating Proplets into Complex Content

The derivation order of the DBS hear mode is *time-linear* by always concatenating a
sentence start and a next word with a semantic relation into a new sentence start:

17.6.1 TIME-LINEAR SURFACE-COMPOSITIONAL HEAR MODE DERIVATION

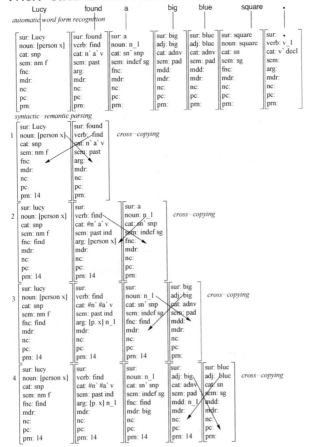

5

sur: lucy	sur:	sur:	sur:	sur:	sur: square	
noun: [person x]	verb: find	noun: n_l	adj: big	adj: blue	noun: square	*absorption*
cat: snp	cat: #n' #a' v	cat: sn' snp	cat: adnv	cat: adnv	cat: sn	*with*
sem: nm f	sem: past ind	sem: indef sg	sem: pad	sem: pad	sem: sg	*simultaneous*
fnc: find	arg: [p. x] n_l	fnc: find	mdd: n_l	mdd:	fnc:	*substitution*
mdr:	mdr:	mdr: big	mdr:	mdr:	mdr:	
nc:	nc:	nc:	nc: blue	nc:	nc:	
pc:	pc:	pc:	pc:	pc: big	pc:	
prn: 14	prn: 14	prn: 14	prn: 14	prn: 14	prn:	

6

sur: lucy	sur:	sur:	sur:	sur:	sur: .	
noun: [person x]	verb: find	noun: square	adj: big	adj: blue	verb: v_1	*absorption*
cat: snp	cat: #n' #a' v	cat: sn' snp	cat: adnv	cat: adnv	cat: v' decl	
sem: nm f	sem: past ind	sem: indef sg	sem: pad	sem: pad	sem:	
fnc: find	arg: [p. x] square	fnc: find	mdd: square	mdd:	arg:	
mdr:	mdr:	mdr: big	mdr:	mdr:	mdr:	
nc:	nc:	nc:	nc: blue	nc:	nc:	
pc:	pc:	pc:	pc:	pc: big	pc:	
prn: 14	prn: 14	prn: 14	prn: 14	prn: 14	prn:	

result

sur: lucy	sur:	sur:	sur:	sur:
noun: [person x]	verb: find	noun: square	adj: big	adj: blue
cat: snp	cat: #n' #a' decl	cat: sn' snp	cat: adnv	cat: adnv
sem: nm f	sem: past ind	sem: indef sg	sem: pad	sem: pad
fnc: find	arg: [p. x] square	fnc: find	mdd: square	mdd:
mdr:	mdr:	mdr: big	mdr:	mdr:
nc:	nc:	nc:	nc: blue	nc:
pc:	pc:	pc:	pc:	pc: big
prn: 14	prn: 14	prn: 14	prn: 14	prn: 14

The hear mode operations use the connectives (i) × for cross-copying (lines 1–4), (ii) ∪ for absorption (line 5), and (iii) ∼ for suspension. Cross-copying encodes the semantic relations of structure such as SBJ×PRED. Absorption combines a function word with a content word such as DET∪CN or with another function word as in PREP∪DET (preposition∪determiner, CLaTR 7.2.5). Suspension such as ADV∼NOM (TExer 3.1.3) applies if no semantic relation exists for connecting the next word with the content processed so far, as in Perhaps ∼ Fido (slept.).

Each derivation step 'consumes' exactly one next word (reading). The language-dependent sur value provided by lexical lookup is omitted in the output.[3] Lexical lookup and syntactic-semantic concatenation are incrementally intertwined: lookup of a new next word occurs only after the current next word has been processed into the current sentence start.[4]

17.7 Speak Mode: Linearization of a Content by Navigation

The speak mode takes a content like 17.5.1 as input and produces a language-dependent surface as output. Graphically, the semantic relations of functor-argument are represented by the connectives / for subject/predicate, \ for object\predicate, and | for modifier|noun, modifier|verb, and modifier|modifier. The semantic relations of coordination are represented graphically by the connective (a) − for noun−noun, (b) verb−verb, (c) adn−adn, and (d) adv−adv.

Based on the definition of graphical /, \, |, and − for the semantic relations of structure, DBS analyzes a content like 17.5.1 in four standard views:

---

[3] A partial exception are name proplets, which preserve their sur value in the form of a marker written in lower case default font, e.g., lucy. In the speak mode, the marker is converted back into a regular sur value written in Helvetica, e.g., Lucy.

[4] The data coverage of DBS is shown in TExer with the explicit definition of 24 linguistically informed examples of English in the hear and the speak mode.

17.7.1 SEMANTIC RELATIONS UNDERLYING SPEAK MODE DERIVATION

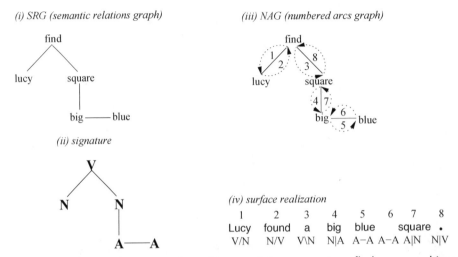

(i) SRG (semantic relations graph)

(iii) NAG (numbered arcs graph)

(ii) signature

(iv) surface realization

1	2	3	4	5	6	7	8
Lucy	found	a	big	blue		square	.
V/N	N/V	V\N	N|A	A–A	A–A	A|N	N|V

The (i) SRG uses the sur marker of lucy and the core values find, square, big and blue of 17.5.1 as nodes. The (ii) *signature* uses the core attributes N(oun), V(erb), and A(dj) as nodes. The (iii) NAG completes the SRG with traversal numbers and shows content activation by the time-linear navigation through the semantic hierarchy in the think mode. The traversal numbers are used in the (iv); it optionally realizes language-dependent surfaces in a speak mode which rides piggyback on the think mode navigation.

In summary, the input to the speak mode is a hierarchical content (17.5.1). The speak mode's time-linear navigation (17.7.1) through the input content achieves a *linearization* of the semantic hierarchy into a sequence of raw surface data as output. The raw data are produced from types by type-token adaptation.

The input to the hear mode is a time-linear sequence of raw surface data. The hear mode's surface-compositional derivation (17.6.1) achieves a re-*hierarchization* into a content; in successful communication, the speaker's input content equals the hearer's output content.

17.8 Natural Language Communication in Speech and Writing

In phylogenetic and ontogenetic evolution, nonlanguage cognition precedes language cognition. In the spirit of Charles Darwin, DBS extends nonlanguage action and recognition to the additional function of language surface production in the speak mode and surface interpretation in the hear mode. Extending the type-token distinction from nonlanguage recognition and action to the hear and speak mode of language cognition may be shown schematically as follows:

17.8.1 EXTENDING NONLANGUAGE INTO LANGUAGE COGNITION

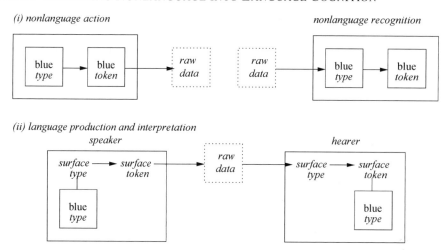

(i) nonlanguage action   nonlanguage recognition

(ii) language production and interpretation
          speaker   hearer

In (i), action and recognition are alike in that they start with the type of the type-token relation. They differ in that the trigger of action is cognition-internal while the trigger of recognition is cognition-external. The output is in complementary distribution, i.e., cognition-external in action and cognition-internal in recognition.

In (ii), action and recognition are moved up to language-dependent surfaces which are connected to content by conventions every speaker-hearer of the language community had to learn (de Saussure 1916, first law: l'arbitraire du signe). As in nonlanguage cognition, production and interpretation of language surfaces have in common that they start with the type of the type-token relation, and differ in that the trigger of the speak mode (production) is cognition-internal while the input to the hear mode (interpretation) is cognition-external. The output is in complementary distribution, i.e., cognition-external in the speak mode and cognition-internal in the hear mode.

Type-token adaptation in speak mode surface production may be illustrated as follows (shown for the medium of writing):

17.8.2 SPEAK MODE: FROM CONTENT TO SURFACE TYPE TO RAW DATA

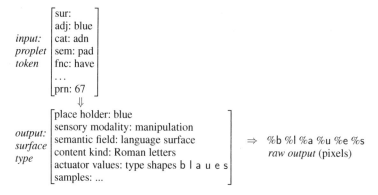

The core value of the proplet token *blue* (content) retrieves the language-dependent surface, here the type of German b l a u e s, based on a list which provides allomorphs using the input proplet's core, cat, and sem values (17.5, category complex). This output serves as input to a realization operation of the agent's interface component which adapts the surface type into a token, realized as raw data.

Type-token recognition in the hear mode may be illustrated as follows:

17.8.3 HEAR MODE: RAW DATA TO SURFACE TYPE TO SURFACE TOKEN

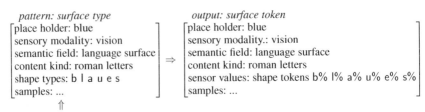

pattern: surface type

$$
\begin{bmatrix}
\text{place holder: blue} \\
\text{sensory modality: vision} \\
\text{semantic field: language surface} \\
\text{content kind: roman letters} \\
\text{shape types: b l a u e s} \\
\text{samples: ...}
\end{bmatrix}
\Rightarrow
$$

output: surface token

$$
\begin{bmatrix}
\text{place holder: blue} \\
\text{sensory modality.: vision} \\
\text{semantic field: language surface} \\
\text{content kind: roman letters} \\
\text{sensor values: shape tokens b\% l\% a\% u\% e\% s\%} \\
\text{samples: ...}
\end{bmatrix}
$$

⇑

raw input (pixels)

The input consists of raw data, provided by the agent's vision sensors and matched by the letters' shape types provided by the agent's memory. The output replaces the shape types, here b l a u e s, with the matching raw data resulting in shape tokens; they are shown as b% l% a% u% e% s% and record the accidental properties. The value crucial for the hearer's understanding, however, is the place holder, here *blue*, for the lexical look-up of the correct nonlanguage concept (17.2.1).

The language dependent surface types, the content types, and the conventions connecting the surface types with the content types exist solely[5] in the respective cognitions of speaker and hearer. This accounts for the fact that for communication to be successful, speaker and hearer must have *learned* the same natural language, including the ability to produce surface types as tokens in the speak mode and recognizing the surface tokens by means of matching types in the hear mode.

17.9 Conclusion

In natural language communication, the transfer of content from speaker to hearer is achieved incrementally by a time-linear sequence of raw data (sound waves in the medium of speech, light waves in writing, etc.) produced as output by the speak mode and serving as input to the hear mode. This constitutes the *language channel* of data-driven agent-based DBS. While (i) navigating the semantic hierarchy in the speak mode (17.7.1) and (ii) reconstructing the semantic hiearchy in the hear mode (17.6.1) have found efficient software solutions in DBS (linear, TCS'92), an operational reconstruction of cultural concepts remains a challenge. In search for a solution, it is proposed to combine the culture-dependent prototypes of Rosch (1973, 1974) with well-established lexical definitions, accommodated by the computational data structure of proplets (17.3), defined as nonrecursive feature structures with ordered attributes.

---

[5] Anything else would be reification, which is uniquely inappropriate for building a talking robot.

18. Paraphrase and Ambiguity

In natural language communication, the speak mode maps cognition-internal content into raw surface data,[1] while the hear mode maps raw surface data into content. The speaker may have a choice between different surfaces for the same content, called *paraphrase*, and the hearer may have to choose between different contents for the same surface, called *ambiguity* (FoCL 11.3).

The restriction of paraphrase to the speak mode and of ambiguity to the hear mode is not reflected in Generative Grammar because its *sign-based substitution-driven* ontology aims at characterizing "well-formedness of expressions," excluding communication (Nativism). For Database Semantics (DBS) with its *agent-based data-driven* ontology, in contrast, the restrictions are fundamental.

18.1 Introduction: the Structure of Content

The definition of ambiguity as a surface representing more than one content and of paraphrase as a content with more than one surface requires a definition of content. In natural language, there are three kinds of elementary contents, defined in DBS as follows:

18.1.1 BASIC BUILDING BLOCKS OF CONTENT

a. concepts
are types which match raw data, resulting in tokens.

b. indexicals
receive their interpretation by pointing at the STAR[2] of an utterance.

c. function words
[3] modify concepts or indexical by taking them as arguments.

In the speak mode, the building blocks are connected by the semantic relations of structure, represented graphically as a / line for subject/predicate, a \ line for predicate\object, a | line for modifier|modified, and a − line for conjunct−conjunct. As an example, consider the following semantic relations analysis of Lucy found a big blue square, underlying the speak mode:

---

[1] I.e. cognition-external sound waves in speech and pixels in writing.

[2] In DBS, the STAR stands for the SPACE, TIME, AGENT (speaker), and RECIPIENT (hearer).

[3] Determiners, conjunctions, and prepositions. In natural language, they may be coded in the morphology (classical Latin) or in the syntax (English).

© The Author(s), under exclusive license to Springer Nature Switzerland AG 2023
R. Hausser, *Ontology of Communication*, https://doi.org/10.1007/978-3-031-22739-4_18

18.1.2 GRAPHICAL REPRESENTATION OF A COMPLEX CONTENT

(i) SRG (semantic relations graph)   *(iii) NAG (numbered arcs graph)*

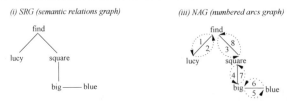

The speak mode is driven by a time-linear navigation along the numbered arcs.

The corresponding hear mode derivation is a surface-compositional time-linear concatenation of proplets (nonrecursive feature structures with ordered attributes[4]):

18.1.3 HEAR MODE CONSTRUCTS CONTENT FROM INPUT SURFACE

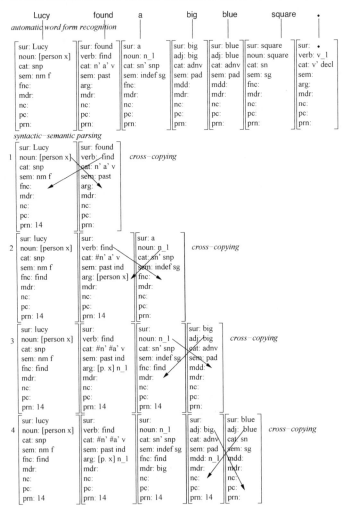

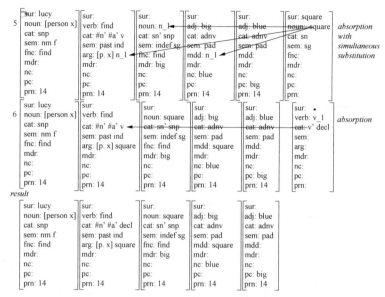

The hear mode operations are (1) crosscopying (connective $\times$), (2) absorption (connective $\cup$), and (3) suspension (connective $\sim$).

Proplets serve as the computational data structure of DBS. The format encodes unlimited grammatical detail, both in terms of proplet attributes and of their values, yet there is no increase of computational complexity above linear because the processing is without recursion or iteration. Instead DBS relies on a strictly time-linear (i.e. left-associative[5]) derivation order in the speak and the hear mode.

18.2 Speak Mode Paraphrase: Different Surfaces for Same Content

A standard example of paraphrase is the active-passive alternation:

18.2.1 PARAPHRASE: DIFFERENT SURFACES FOR A SINGLE CONTENT

Mary read a book.
A book was read by Mary.

As a speak mode phenomenon, DBS analyzes paraphrase as different traversals of the same content. For example, the common content of the paraphrases 18.2.1 is represented as the following numbered arcs graph (NAG):

---

[4] Our definition is the direct opposite to *recursive* feature structures with *unordered* attributes (Carpenter 1992). Popular in Generative Grammar, recursive feature structures with unordered attributes are maximally inefficient for computational pattern matching, but justified by a misled notion of generality.

[5] Aho and Ullman 1977, p.47.

18.2.2 USING DIFFERENT TRAVERSAL ORDERS IN THE SPEAK MODE

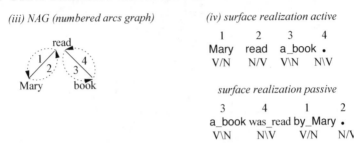

(iii) NAG (numbered arcs graph)

(iv) surface realization active

1	2	3	4
Mary	read	a_book	.
V/N	N/V	V\N	N\V

surface realization passive

3	4	1	2
a_book	was_read	by_Mary	.
V\N	N\V	V/N	N/V

Each line in a semantic relations graph has a forward (downward V/N or V\N) and a backward (upward, V/N or V\N) traversal, indicated by dotted arrows in the NAG (CLaTR 7.4.1). Traversals may be empty, i.e. without a surface realization[6]. Like the hear mode (18.1.3), the speak mode (18.2.2) is strictly *surface-compositional* and *time-linear* (methodological principles of DBS).

18.3 DBS Formalism for the Speak Mode (Language Production)

In the active variant of 18.2.2, the first rule to apply in the sequence of speak mode operations is V/N:

18.3.1 NAVIGATING WITH V/N FROM *read* TO *Mary* (arc 1)

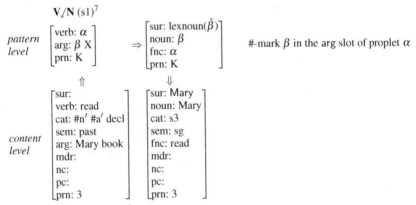

The operation lexnoun($\hat{\beta}$) in the sur slot of the goal proplet realizes the English surface Mary, based on β matching the initial value of the arg slot in the input proplet.

To realize the predicate and acquire the filler of the object slot, the navigation returns to the V with the speak mode operation N/V:

---

[6] For example the traversal of arc 6 and 7 in 18.1.2.

[7] The 's' indicates a speak mode operation and the '1' refers to the operation number in the DBS speak mode grammar defined in TExer 6.5.1.

18.3.2 NAVIGATING WITH N∕V FROM *Mary* TO *read* (arc 2)

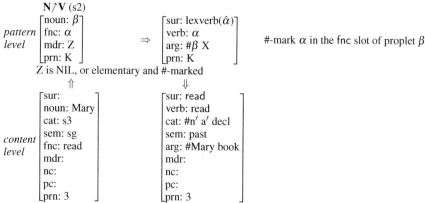

The #-marking of the first arg value in the goal proplet resulted from the instruction of V∕N (18.3.1). A #-marking instruction applies to a feature, here [arg: #Mary book], and not just to the value. For example, if a value in an arg slot is being #-marked, this does not affect the same value in a mdd slot.

Continuing the navigation from the predicate to the object is based on V\N:

18.3.3 NAVIGATING WITH V\N FROM *read* TO *book* (arc 3)

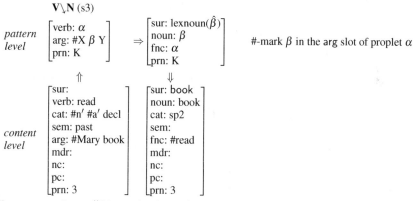

The arg value #Mary is bound to the variable #X in the input pattern [arg: #X β Y], the arg value book to β, and a possible third argument (in a three place verb) would be bound to Y.

The return from the object to the predicate with N↑V is motivated by the need (i) to realize the punctuation mark (period), and (ii) to get into position for navigating to a possible successor proposition:

18.3.4 NAVIGATING WITH N↑\V FROM *book* BACK TO *read* (arc 4)

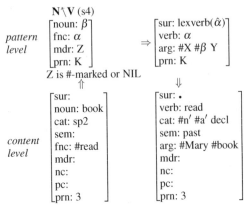

$$\begin{array}{l} \textbf{N}\backslash\textbf{V} \text{ (s4)} \\ \textit{pattern} \\ \textit{level} \end{array} \begin{bmatrix} \text{noun: } \beta \\ \text{fnc: } \alpha \\ \text{mdr: Z} \\ \text{prn: K} \end{bmatrix} \Rightarrow \begin{bmatrix} \text{sur: lexverb}(\hat{\alpha}) \\ \text{verb: } \alpha \\ \text{arg: \#X \#}\beta \text{ Y} \\ \text{prn: K} \end{bmatrix}$$

Z is #-marked or NIL

$$\begin{array}{l} \textit{content} \\ \textit{level} \end{array} \begin{bmatrix} \text{sur:} \\ \text{noun: book} \\ \text{cat: sp2} \\ \text{sem:} \\ \text{fnc: \#read} \\ \text{mdr:} \\ \text{nc:} \\ \text{pc:} \\ \text{prn: 3} \end{bmatrix} \begin{bmatrix} \text{sur: } \bullet \\ \text{verb: read} \\ \text{cat: \#n' \#a' decl} \\ \text{sem: past} \\ \text{arg: \#Mary \#book} \\ \text{mdr:} \\ \text{nc:} \\ \text{pc:} \\ \text{prn: 3} \end{bmatrix}$$

The traversal of the corresponding passive paraphrase in 18.2.2 uses the same speak mode operations, but in the order V\N, N↑\V, V↙N, N↗V.

18.4 Hear Mode Ambiguity: Different Contents for Same Surface

Syntactic ambiguity (FoCL Sect. 2.5) as a language-dependent hear mode phenomenon is a single surface for more than one content. A classic example in English is the alternative between the adnominal and adverbial use of a modifier.

18.4.1 FIRST READING: ADNOMINAL USE OF A MODIFIER

(iii) NAG (numbered arcs graph)

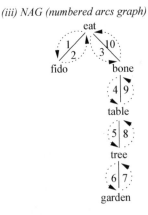

(iv) surface realization

1	2	3	4	5	6	7	8	9	10
Fido	ate	the_bone	on_the_table	under_the_tree	in_the_garden				
V/N	N/V	V\N	N|N	N|N	N|N	N|N	N|N	N|N	N|V

In this reading, ON THE TABLE IN THE GARDEN UNDER THE TREE modifies BONE as an adnominal.

18.4.2 SECOND READING: ADVERBIAL USE

(iii) NAG (numbered arcs graph)

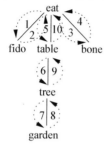

(iv) surface realization

1	2	3	4	5	6	7	8	9	10
Fido	ate	the_bone		on_the_table	under_the_tree	in_the_garden			.
V/N	N/V	V\N	N\V	N|N	N|N	N|N	N|N	N|N	N\V

In this reading, ON THE TABLE IN THE GARDEN UNDER THE TREE modifies EAT as an adverbial.

18.5 Ambiguity is Language-Dependent

Ambiguities, lexical as well as syntactic, are language-dependent. For example, the following example is syntactically ambiguous in English, but its translation into German has two separate unambiguous readings:

18.5.1 FLYING AIRPLANES CAN BE DANGEROUS.

(a) Fliegende Flugzeuge können gefährlich sein.
(b) Flugzeuge zu fliegen kann gefährlich sein.

The cause of this ambiguity is the absence of a morphological distinction between the adnominal (fliegende) and the infinitival (zu fliegen) use of the English participle flying.

Another syntactic ambiguity in English with an unambiguous counterpart in German is the following:

18.5.2 THEY DON'T KNOW HOW GOOD MEAT TASTES.

(a) Sie wissen nicht wie gut Fleisch schmeckt.
(b) Sie wissen nicht wie gutes Fleisch schmeckt.

The cause of this ambiguity is the absence of a morphological distinction between the adverbial (gut) and adnominal (gutes) use of English good.

18.6 Grammatical Analysis of Ambiguity

Because the ambiguity of Flying airplanes can be dangerous is syntactic, it is treated in DBS by alternative syntactic-semantic operations, called **PROG×NP** and **ADN×PN**. These two hear mode operations take the same input, defined as follows:

18.6.1 LEXICAL ANALYSIS OF FLYING (NLC A5.1.5) AND AIRPLANES:

On one reading, the time-linear concatenation combines flying with a noun phrase as the object-completion of a reduced infinitive:

18.6.2 CROSS-COPYING FLYING AND AIRPLANES IN OBJ COMPLETION

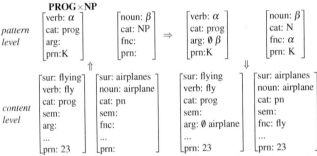

This hear mode operation connects the proplets *flying* and *airplanes* by cross-copying between the core value [verb: fly] into the continuation slot [fnc:] of *airplanes* and the core value [noun: airplane] into the [arg:] slot of *flying*.

The other reading combines a modifier with a plural noun. This constellation may be with and without a determiner.[8]

18.6.3 CROSS-COPYING FLYING AND AIRPLANES IN PN-MODIFICATION

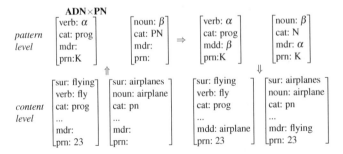

The two readings differ in that the cross-copying in 18.6.2 is from the core features [verb: fly] and [noun: airplane] into the continuation slots fnc and arg (completing the transitive infinitive with an object), but in 18.6.3 from the core features into the continuation slots mdd and mdr (modifying airplanes with flying).

18.7 Local vs. Global Ambiguities

The ambiguities in Sect. 18.5 are called [+global][9] in DBS because they hold for complete expressions (sentence, proposition). An example of a [-global] or local ambiguity, in contrast, is the famous 'Gardenpath'[10] sentence by Bever (1970):

18.7.1 Local Ambiguity

Gardenpath sentence: THE HORSE RACED BY THE BARN (a) .

(b) FELL.

The continuation horse+raced introduces a [-global] ambiguity between (a) *horse raced* (active) and (b) *horse which was raced* (passive), resulting in two parallel derivation strands up to and including barn. Depending on continuing after barn with (a) an interpunctuation or (b) a verb, one of the two [-global] readings is grammatically disambiguated.

18.8 Iterating Local Ambiguities

Local ambiguities may be iterated, as shown by the following examples:

18.8.1 OBJECT-CLAUSE ITERATION

A. Bob believes Bill.
B. Bob believes that Bill believes Mary.
C. Bob believes that Bill believes that Mary believes Suzy.
D. Bob believes that Bill believes that Mary believes that Suzy believes Tim.

Here, the local ambiguities are between concluding with a full stop and continuing with a subclause. Local ambiguities do not affect the linear time complexity of natural language grammars in DBS.

---

[8] E.g. THE FLYING AIRPLANES VS. FLYING AIRPLANES. For comparison with 18.6.2, the operation application 18.6.3 shows the latter.

[9] The ±global distinction between ambiguities presupposes a time-linear interpretation of natural language, i.e. the computation of *possible continuations*, as in surface-compositional, time-linear DBS, which is agent-based data-driven. This is in contradistinction to sign-based substitution-driven Phrase Structure Grammar, which computes *possible substitutions* to characterize well-formedness without the distinction between the speak and the hear mode, i.e. regardless of communication (Nativism).

[10] So called because the initial interpretation up to barn is misleading, as in 'leading someone down the garden path'. In an era of substitution-driven "Generative Grammar," Bever's example is wide awake and far ahead of its time.

18.9 Conclusion

The limitation of paraphrase to the speak mode and of ambiguity to the hear mode is a general phenomenon of natural language. When building a talking robot, the processing of paraphrase must be built only for the speak mode and the processing of ambiguity, local or global, must be built only for the hear mode.

Paraphrase of the speak mode does not affect the computational complexity of natural language communication. The only possible source of a complexity degree above linear would be ambiguity, restricted to the hear mode. However, to affect complexity, ambiguity would have to be (i) iterative/recursive and (ii) at least two readings would have to 'survive' each cycle. As shown by 13.9.1, (ii) is not the case.

19. Recursion and Grammatical Disambiguation

The grammatical structure of natural language signs includes recursion. This raises the question of whether unbounded recursion may cause an unbounded increase of ambiguity. If true, it would be an obstacle to successful communication. Also the construction of a linguistic example is most challenging.

Moreover, there is empirical evidence that natural language prevents systematic ambiguity by means of systematic grammatical disambiguation. As a general phenomenon, grammatical disambiguation is a candidate for being a natural language universal.

Database Semantics is a linguistic theory of natural language communication which reconstructs a content-surface mapping in the *speak mode* and a surface-content mapping in the *hear mode*.[1] Thereby, the speak mode may have to choose between paraphrases (e.g. active vs. passive) and the hear mode between the readings of an ambiguity (e.g. adnominal vs. adverbial modification).

19.1 Speak Mode in Database Semantics

The speak mode is the language variant of an agent's *action*. The input is a content, the output a language-dependent surface. The drive or motor of the speak mode is a navigation along the semantic relations in the input content. Consider the following speak mode analysis of LUCY FOUND A BIG BLUE SQUARE:

19.1.1 GRAPHICAL DBS ANALYSIS OF A SPEAK MODE EXAMPLE

(i) SRG (semantic relations graph)   *(iii) NAG (numbered arcs graph)*

The semantic relations of structure are subject/predicate, object\predicate, modifier|modified, and conjunct−conjunct (Hausser 2022a).

---

[1] This is in contradistinction to Generative Grammar (Chomsky 1957, 1965), which aims to randomly generate all well-formed expressions of a natural language from a single abstract input expression, called the S node (Nativism). The ontology of Generative Grammar is sign-based substitution-driven, that of Database Semantics agent-based data-driven.

R. Hausser, *Ontology of Communication*, https://doi.org/10.1007/978-3-031-22739-4_19

19.2 Hear Mode in Database Semantics

The hear mode is the language variant of an agent's *recognition*. The input is a language-dependent surface and the output a content. The drive or motor is a sequence of word forms in the form of raw data which are recognized by type-token matching in the agent's interface component (Hausser 2021, 12.8.1):

19.2.1 SURFACE-COMPOSITIONAL TIME-LINEAR HEAR MODE DERIVATION

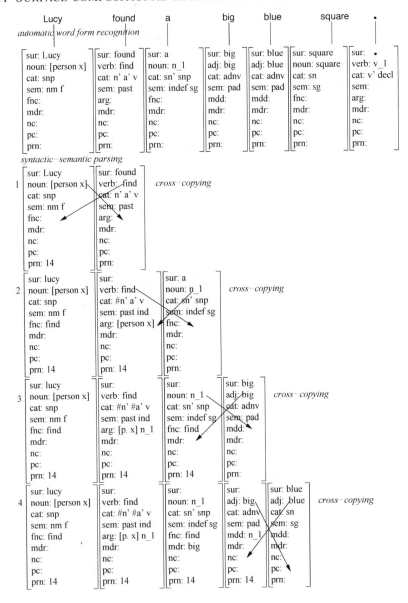

```
     ┌sur: lucy        ┐┌sur:              ┐┌sur:            ┐┌sur:          ┐┌sur:         ┐┌sur: square   ┐
   5 │noun: [person x] ││verb: find        ││noun: n_1       ││adj: big      ││adj: blue    ││noun: square  │  absorption
     │cat: snp         ││cat: #n' #a' v    ││cat: sn' snp    ││cat: adnv     ││cat: adnv    ││cat: sn       │  with
     │sem: nm f        ││sem: past ind     ││sem: indef sg   ││sem: pad      ││sem: pad     ││sem: sg       │  simultaneous
     │fnc: find        ││arg: [p. x] n_1   ││fnc: find       ││mdd: n_1      ││mdd:         ││fnc:          │  substitution
     │mdr:             ││mdr:              ││mdr: big        ││mdr:          ││mdr:         ││mdr:          │
     │nc:              ││nc:               ││nc:             ││nc: blue      ││nc:          ││nc:           │
     │pc:              ││pc:               ││pc:             ││pc:           ││pc: big      ││pc:           │
     └prn: 14          ┘└prn: 14           ┘└prn: 14         ┘└prn: 14       ┘└prn: 14      ┘└prn:          ┘

     ┌sur: lucy        ┐┌sur:              ┐┌sur:            ┐┌sur:          ┐┌sur:         ┐┌sur: •        ┐
   6 │noun: [person x] ││verb: find        ││noun: square    ││adj: big      ││adj: blue    ││verb: v_1     │  absorption
     │cat: snp         ││cat: #n' #a' v    ││cat: sn' snp    ││cat: adnv     ││cat: adnv    ││cat: v' decl  │
     │sem: nm f        ││sem: past ind     ││sem: indef sg   ││sem: pad      ││sem: pad     ││sem:          │
     │fnc: find        ││arg: [p. x] square││fnc: find       ││mdd: square   ││mdd:         ││arg:          │
     │mdr:             ││mdr:              ││mdr: big        ││mdr:          ││mdr:         ││mdr:          │
     │nc:              ││nc:               ││nc:             ││nc: blue      ││nc:          ││nc:           │
     │pc:              ││pc:               ││pc:             ││pc:           ││pc: big      ││pc:           │
     └prn: 14          ┘└prn: 14           ┘└prn: 14         ┘└prn: 14       ┘└prn: 14      ┘└prn:          ┘
result

     ┌sur: lucy        ┐┌sur:              ┐┌sur:            ┐┌sur:          ┐┌sur:         ┐
     │noun: [person x] ││verb: find        ││noun: square    ││adj: big      ││adj: blue    │
     │cat: snp         ││cat: #n' #a' decl ││cat: sn' snp    ││cat: adnv     ││cat: adnv    │
     │sem: nm f        ││sem: past ind     ││sem: indef sg   ││sem: pad      ││sem: pad     │
     │fnc: find        ││arg: [p. x] square││fnc: find       ││mdd: square   ││mdd:         │
     │mdr:             ││mdr:              ││mdr: big        ││mdr:          ││mdr:         │
     │nc:              ││nc:               ││nc:             ││nc: blue      ││nc:          │
     │pc:              ││pc:               ││pc:             ││pc:           ││pc: big      │
     └prn: 14          ┘└prn: 14           ┘└prn: 14         ┘└prn: 14       ┘└prn: 14      ┘
```

The connectives of the hear mode are × (cross-copying), ∪ (absorption), and ∼ (suspension, TExer Sect. 8.3).

The building blocks of a content as input to the speak mode and output of the hear mode are nonrecursive feature structures with ordered attributes,[2] called *proplets*. By coding the semantic relations in a content by address, the proplets of a content are order-free. This is essential for the storage and retrieval in the content-addressable database of DBS[3] of artificial cognition (NLC 3.3).

19.3 Recursion

The following example combines recursion and ambiguity:

19.3.1 COMBINING RECURSION AND AMBIGUITY

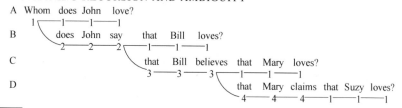

---

[2] This is the direct opposite to the feature structures popular in Generative Grammar, which are *recursive* with *unordered* attributes (Carpenter 1992). Motivated by a misguided notion of generality, recursive feature structures with unordered attributes are maximally inefficient for computational pattern matching and superfluous in agent-based data-driven DBS.

[3] Called 'A-memory', earlier 'word bank'.

The ambiguities arise in continuing after John with a verb taking either a noun or a clause as object. For example, John is ambiguous between between terminating with loves Mary or continuing with said that. The ambiguities creating the recursion are strictly local ([-GLOBAL], FocL 13.3.6).

Because each complete line in 19.1.1 is unambiguous, the systematic ambiguity originating in this recursion does not affect the linear time complexity of natural language (Hausser 2022b). Furthermore, systematic grammatical disambiguation in recursion holds for natural language in general (universal).

The drive (motor) of the speak mode is a navigation along the semantic relations in a content, called *coactivation*:[4]

19.3.2 COACTIVATION BY THE TRAVERSAL OF COGNITIVE CONTENT

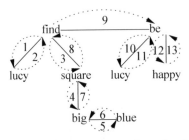

The drive of the hear mode, on the other hand, is the sequence of incoming language-dependent surfaces, as illustrated by the following example:

19.3.3 TIME-LINEAR SURFACE-COMPOSITIONAL DERIVATION

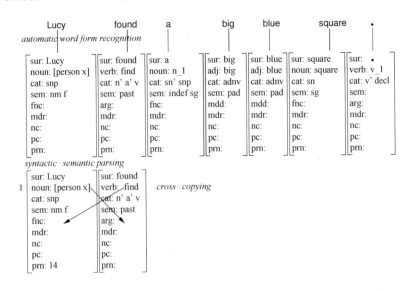

---

[4] In the sense of driving surface production by a traversal of cognitive content (thought).

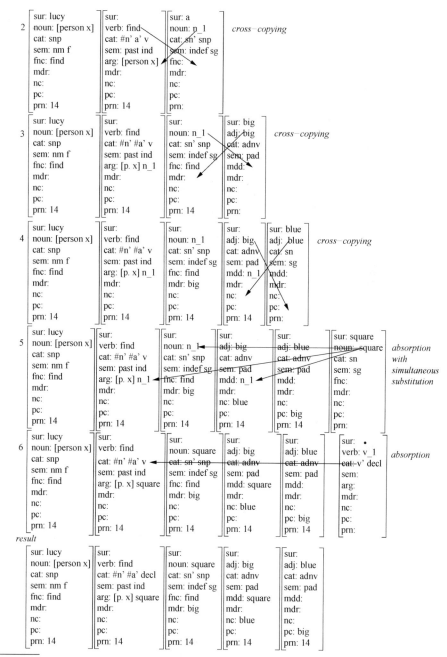

<hr />

[4] An example of a [-global] ambiguity is the famous 'Gardenpath' sentence "THE HORSE RACED BY THE BARN FELL" by Bever (1970), so-called because the initial interpretation up to barn is misleading, as in 'leading someone down the garden path'. In an era of substitution-driven "Generative Grammar", Bever's continuation-based example is wide awake and far ahead of its time.

The derivation satisfies the two methodological principles of DBS: (i) *surface-compositionality*[5] and (ii) *time-linearity*[6].

19.4 Conclusion

For a theoretical claim to be a linguistic universal, it must be without a counterexample. Empirically, this is hard to prove. For example, even if a broad-based investigation of unbounded ambiguity in an unbounded recursion does not come up with a clear example, the claim of nonexistence can not be conclusive because there remains the *possibility* of having overlooked such an instance. There remains, however, the functional argument: unbounded ambiguity with unbounded recursion is unlikely because it would be an obstacle to successful communication.

---

[5] "The analysis of natural language signs is surface-compositional if it uses concrete word forms as building blocks, such that syntactic and semantic properties of a complex expression derive systematically from (i) the lexical properties of the related word forms and (ii) their standard syntactic-semantic composition" (FoCL 4.5.1).

[6] Left-associative derivation order, Aho & Ullman 1977, p.47.

Name Index

Bibliography

Aho, A.V., B.W. Kernighan, and P.J. Weinberger (1977, 1988) *The AWK Programming Language*, Addison-Wesley

Aho, A.V. and J.D. Ullman (1977) *Principles of Compiler Design*, Reading, Mass.: Addison-Wesley

Austin, J.L. ([1955]1962) *How to Do Things with Words*, Oxford: Clarendon Press

Aristotle (1974). *The Complete Works of Aristotle*, edited by Jonathan Barnes, 2 vols., Princeton University Press

Ballmer, T. (1978) *Logical Grammar*, Amsterdam: North Holland

Bar-Hillel, Y. (1964) *Language and Information. Selected Essays on Their Theory and Application*, Reading, Mass.: Addison-Wesley

Barsalou, W., W.K. Simmons, A.K. Barbey, and C.D. Wilson (2003)"Grounding conceptual knowledge in modality-specific systems," *TRENDS in Cognitive Sciences*, Elsevier, Vol. 7.2:84–91

Barwise, J., and J. Perry (1983) *Situations and Attitudes*, Cambridge, Mass.: MIT Press

Bayer, S. (1996) "The coordination of unlike categories," *Language*, Vol. 72:579–616

Benson, D.F. (1994) *The Neurology of Thinking*, New York: Oxford University Press (OUP) *https://doi.org/10.1002/ana.410360535*

Berwick, R.C., and A.S. Weinberg (1984) *The Grammatical Basis of Linguistic Performance: Language Use and Acquisition*, Cambridge, Mass.: MIT Press

Barcan Marcus, R. (1961) "Modalities and Intensional Languages," *Synthese*, Vol. 13.4:303–322

Bever, T.G (1970) "The cognitive basis for linguistic structures". In: J.R. Hayes (ed.) *Cognition and the development of language*, pp. 279-362, New York: Wiley

Biederman, I. (1987) "Recognition-by-components: a theory of human image understanding," *Psychological Review*, Vol. 94:115–147

Bjørner, D. (1978) *Programming in the Meta-Language: A Tutorial*, in LNCS 61:337–374, Springer

Bloomfield, L. (1933) *Language*, New York: Holt, Rinehart, and Winston

Boas, F. (1911) *Handbook of American Indian languages (Vol. 1)*, Bureau of American Ethnology, Bulletin 40. Washington: Government Print Office (Smithsonian Institution, Bureau of American Ethnology)

Briandais, R. de la (1959) "File searching using variable length keys," *Proc. Western Joint Computer Conf. 15*, 295–298

Brooks, R.A. (1986) "A Robust Layered Control System for a Mobile Robot," *IEEE Journal of Robotics and Automation*, Vol. 2.1:14–23

Brown, R. (1958) "How shall a thing be called?" *Philosophical Review*, Vol. 65.1:14-21, Duke University Press

Brüning, B., and E. Al Khalaf (2020) "Category mismatches in coordination revisited," *Linguistic Inquiry*, Vol. 31.1:1–36

Bunt, H., J. Thesingh, and K. van der Sloot (1987) "Discontinuous Constituents in Trees, Rules, and Parsing," in *Proc. 3rd Conference of the European chapter of the ACL*, pp. 203-210. Copenhagen, DK, Univ. of Copenhagen

Carbonell, J.G. and R. Joseph (1986) "FrameKit$^+$: a knowledge representation system," Carnegie Mellon University, Department of Computer Science

Carpenter, B. (1992) *The Logic of Typed Feature Structures*, Cambridge: Cambridge University Press

Chisvin, L., and R.J. Duckworth (1992) "Content-addressable and associative memory," pp. 159–235, in M.C. Yovits (ed.) *Advances in Computer Science*, 2nd ed., Academic Press

Chomsky, N. (1965) *Aspects of the Theory of Syntax*, Cambridge, Mass.: MIT Press

Chomsky, N. (1982) *Language and the Study of Mind*, Cambridge, Mass.: MIT Press

Chomsky, N., and M. Halle (1968) *The Sound Pattern of English*, New York: Harper & Row.

Church, A. (1932). "A set of postulates for the foundation of blogic," *Annals of Mathematics* Series 2. 33 (2): 346–366

Church, K.W., & R.L. Mercer, (1993) "Introduction to the Special Issue on Computational Linguistics Using Large Corpora," ACL, Vol. 19.1:1-24

Cocke, J. and Schwartz, J. T. (1970). Programming languages and their compilers: Preliminary notes (PDF) (Technical report) (2nd revised ed.). CIMS, NYU.

Connell, L., D. Lynott and B. Banks (2017) "Interoception: The forgotten modality in perceptual grounding of abstract and concrete *http://dx.doi.org/10.1098/rstb.2017.0143*

Cresswell, M. (1972) "The world is everything that is the case," *Australasian Journal of Philosophy*, Vol. 50:1–13.

Culicover, P. and R. Jackendoff (1997) "Semantic subordination despite syntactic coordination," *Linguistic Inquiry*, Vol. 28.2:195–217

Darwin, C. (1859) *On the Origin of the Species*, London: John Murray

Delshad, F. (2009) *Georgica et Irano-Semitica, Philologische Studien zu den iranischen und semitischen Elementen im georgischen Nationalepos 'Der Recke im Pantherfell'*, Schriften zur Literaturwissenschaft, Wiesbaden

Dik, S. (1968) *Coordination: Its Implications for a Theory of General Linguistics*, Amsterdam: North Holland Publishing Company

Dodt, E., and Y. Zotterman (1952) "The Discharge of Specific, Cold Fibres at High Temperatures," *Acta Physiologica Scandinavica*, Vol 26.4: 358–365, December 1952

Dretske, F. (1981) *Knowledge and the Flow of Information*, Cambridge, Mass.: Bradford Books/MIT Press

Earley, J. (1970) "An Efficient Context-Free Parsing Algorithm," *Commun. ACM* 13.2:94–102

Ekman, P. (1999) "Basic Emotions," Chapter 3 in *Handbook of cognition and emotion*, T. Dalgleish and M. Powers (eds), John Wileys

Elmasri, R., and S.B. Navathe (2017) *Fundamentals of Database Systems, 7th ed.*, Redwood City, CA: Benjamin-Cummings

Euler, L. (1761) *Lettres a Une Princesse d'Allemagne*, Vol. 2: Letters No. 102–108. St. Petersburg: Imperial Academy of Sciences, 1768–1772

Filingeri, D. (2016) "Neurophysiology of Skin Thermal Sensations," *Comprehensive Physiology* Vol. 6.3:2–78

Fillmore, C.J. (1988) "The Mechanisms of Construction Grammar" *Proceedings of the Fourteenth Annual Meeting of the Berkeley Linguistics Society (1988)*, pp. 35–55

Fischer, W. (2002) *Implementing Database Semantics as an RDBMS* (in German), Studienarbeit am Institut für Informatik der Universität Erlangen-Nürnberg (Prof. Meyer-Wegener), published as CLUE-Arbeitsbericht 7 (2004)

Fredkin, E. (1960) "Trie Memory," *Commun. ACM* Vol. 3.9:490–499

Frege, G. (1879) *Begriffsschrift. Eine der arithmetischen nachgebildete Formelsprache des reinen Denkens*, Halle: L. Nebert

Garey, M.R., and D.S. Johnson (1979) *Computers and Intractability: A Guide to the Theory of NP-Completeness*, San Francisco: W. H. Freeman

Gazdar, G., I. Sag, T. Wasow & S. Weisler (1985) "Coordination and how to distinguish categories," in *Natural Language and Linguistic Theory 3*

Geach, P. (1972) "A Program for Syntax," in D. Davidson and G. Harman (eds.) *Semantics of Natural Language*, Dordrecht: D. Reidel, pp. 483–497

Ginsburg, S. (1980) "Formal Language Theory: Methods for Specifying Formal Languages – Past, Present, Future," in R.V. Book (ed.), 1–22

Goldsmith, J. (1985) "A principled exception of the coordinate structure constraint," *CLS 21*

Gödel, K. (1930), "Die Vollständigkeit der Axiome des logischen Funktionenkalküls," Monatshefte für Mathematik und Physik 37: 349–60

Gödel, K. (1931), "Über die formal unentscheidbare Sätze der Principia Mathematica und verwandter Systeme, I." Monatshefte für Mathematik und Physik 38: 173–98.

Gödel, K. (1932), "Zum intuitionistischen Aussagenkalkül," Anzeiger Akademie der Wissenschaften Wien 69: 65–66

Greibach, S. (1973) "The hardest context-free language," *SIAM J. Comput.* Vol. 2:304–310

Grice, P. (1957) "Meaning," *Philosophical Review*, Vol. 66:377–388

Gualterus Burlaeus (1988) *De puritate artis logicae tractatus longior*, Hamburg: Felix Meiner Verlag

Haddad, R., A. Medhanie, Y. Roth, D. Harel, N. Sobel (2010) "Predicting Odor Pleasantness with an Electronic Nose," *PLoS Comput Biol* Vol. 6.4: e1000740

Handl, J. (2008) *Entwurf und Implementierung einer abstrakten Maschine für die oberflächenkompositionale inkrementelle Analyse natürlicher Sprache*, Diplom Thesis, Department of Computer Science, U. of Erlangen-Nürnberg

Handl, J. (2012) *Inkrementelle Oberflächenkompositionale Analyse und Generierung Natürlicher Sprache*, Inaugural Dissertation, CLUE, Univ. Erlangen–Nürnberg. http://opus4.kobv.de/opus4-fau/frontdoor/indexx/index/doc Id/3933

Hankamer, J. (1979) *Deletion in Coordinate Structures*, New York: Garland Publishing

Harbsmeier, C. (2001) "May Fourth Linguistic Orthodoxy and Rhetoric" *New Terms for New Ideas*, pp. 373–410, ed. Lackner, Kurtz, Amelung, Leiden: BrillHorn, L., and G. Ward (eds.) (2004) *The Handbook of Pragmatics*, Oxford: Blackwell Publishers

Harman, G. (1963) "Generative Grammar without Transformational Rules: a Defense of Phrase Structure," *Language*, Vol. 39:597–616

Harris, Z. [1941](1951) *Methods in Structural Linguistics*, Chicago: Univ. of Chicago

Harrison, M. (1978) *Introduction to Formal Language Theory*, Reading, Mass.: Addison-Wesley

Hartmann, K. (2000) *Right Node Raising and Gapping: Interface Conditions on Prosodic Deletion*, Amsterdam: John Benjamins

Hausser, R. (1980) "The Place of Pragmatics in Model-Theory," in Groenendijk, J.A.G., T.M.V. Janssen, and M.B.J. Stokhof (eds) *Formal Methods in the Study of Language*, University of Amsterdam: Mathematical Center Tracts 135

Hausser, R. (1986) *NEWCAT: Natural Language Parsing Using Left-Associative Grammar*, (Lecture Notes in Computer Science 231), 540 pp., Springer = NEWCAT

Hausser, R. (1989) "Database Semantics for Natural Language." *Artificial Intelligence*, Vol. 106:283-305, Amsterdam: Elsevier = AIJ'89

Hausser, R. ([1989] 2013) *Computation of Language*, Springer = CoL

Hausser, R. (1992) "Complexity in Left-Associative Grammar," *Theoretical Computer Science*, Vol. 106.2:283-308, Amsterdam: Elsevier = TCS'92

Hausser, R., ([1999, 2001] 2014) *Foundations of Computational Linguistics; Human-Computer Communication in Natural Language*, pp. 518. Springer = FoCL

Hausser, R. (2020) *Twentyfour Exercises in Linguistic Analysis, DBS software design for the Hear and the Speak mode of a Talking Robot*, lagrammar.net = TExer

Hausser, R. (2006) *A Computational Model of Natural Language Communication – Interpretation, Inferencing, and Production in Database Semantics*, Springer, pp. 360; preprint 2nd Ed. 2017, pp. 363, at lagrammar.net = NLC

Hausser, R. (2011) *Computational Linguistics and Talking Robots; Processing Content in DBS*, pp. 286. Springer = CLaTR

Hausser, R. (2017) "A computational treatment of generalized reference," *Complex Adaptive Systems Modeling*, Vol. 5.1:1–26, Springer = CASM'17

Hausser, R. (2019) *Computational Cognition, Integrated DBS Software Design for Data-Driven Cognitive Processing*, pp. i–xii, 1–237, lagrammar.net = CC

Hausser, R. (2021a) "Computational Pragmatics," lagrammar.net

Hausser, R. (2021b) "The Grounding of Concepts in Science," lagrammar.net

Hausser, R. (2021c) "Agent-Based Memory as an On-Board Database," lagrammar.net

Hausser, R. (2021d) "Comparison of Coordination and Gapping," lagrammar.net

Hausser, R. (2022) "Grammatical Disambiguation: The natural language linear complexity hypothesis," *Language and Information*, to appear

Hopcroft, J.E., and Ullman, J.D. (1979) *Introduction to Automata Theory, Languages, and Computation*, Reading, Mass.: Addison-Wesley

Hubel, D.H., and T.N. Wiesel (1962) "Receptive Fields, Binocular Interaction, and Functional Architecture in the Cat's Visual Cortex," *Journal of Physiology*, Vol. 160:106–154

Hume, D. ([1739] 1896) *A Treatise of Human Nature*, reprinted from the Original Edition in three volumes and edited, with an analytical index, by L.A. Selby-Bigge, Oxford: Clarendon Press

Hume, D. (1748) *An Enquiry Concerning Human Understanding*, reprinted in Taylor, R. (1974) *The Empiricists: Locke, Berkeley, Hume*, Garden City, New York: Anchor Books, Doubleday

Jackendoff, R. (1971) "Gapping and related rules," *Linguistic Inquiry*, Vol. 2: 21–35

Johansson, I. (2012) "Hume's Ontology," *Metaphysica*, Vol. 13.1:87-105

Johnson, K. (2009) "Gapping is not (VP) ellipsis," *Linguistic Inquiry*, Vol. 40:289–328

Kamp, J.A.W. (1980) "A Theory of Truth and Semantic Representation," in J.A.G. Groenendijk et al. (eds.)

Kamp, J.A.W., and U. Reyle (1993) *From Discourse to Logic*, Parts 1 and 2, Dordrecht: Kluwer

Kant, E. (1783) *Prolegomena*, Reprint of the 6th ed., Leipzig, AA 04

Kasami, T. (1965) "An Efficient Recognition and Syntax-Analysis Algorithm for Context-Free Languages," Air Force Cambridge Research Lab, Bedford, Scientific report AFCRL-65-758

Kempson, R. (2001) "Pragmatics: Language and Communication," Chapter 16 in *The Handbook of Linguistics*, M. Aronoff & J. Rees-Miller (eds.), Oxford: Blackwell Publishers, pp. 394–427

Kiefer, F. (2018) "Two kinds of epistemic modality in Hungarian," in Guentchéva, Z. (ed.) *Empirical Approaches to Language Typology*, DOI: 10.1515/9783110572261–013

King, J.C., and K.S. Lewis, "Anaphora," *The Stanford Encyclopedia of Philosophy (Summer 2017 Edition)*, Edward N. Zalta (ed.), *https://plato.stanford.edu/archives/sum2017/entries/anaphora/*

Kirkpatrick, K. (2001), "Object Recognition", in *Avian Visual Cognition*, cyberbook in cooperation with Comparative Cognitive Press

Kleene, S.C. (1952) INTRODUCTION TO METAMATHEMATICS, Amsterdam
Knuth, D.E., J.H. Morris, and V.R. Pratt (1977) "Fast Pattern Matching in Strings," *SIAM Journal of Computing*, Vol. 6.2:323–350

Kondo, K. (2006) "Post-infectious fatigue" *JMAJ* Vol. 49.1:27–33

Kuno, S. (1976) "Gapping: A functional analysis," *Linguistic Inquiry*, Vol. 7:300–318

Lakoff, G. (1986) "Frame semantic control of the coordinate structure constraint," *CLS 22*

Kycia, A. (2004) *An Implementation of Database Semantics in Java*. M.A. thesis, CLUE

Lerner, J.S., Ye Li, P. Valdesolo, and K.S. Kassam (2015) "Emotion and Decision Making," *Annual Review of Psychology*, Vol. 66:799–823

Levelt, W.J.M. (1981) "The speaker's linearization problem," Transactions of the Royal Society, 295.1077:305-315

Levin, B. (2009) "Where Do Verb Classes Come From?" *http://web.stan ford.edu/bclevin/ghent09vclass.pdf*

Lewis, D. (1969) *Convention: a Philosophical Study*, Hoboken, N.J.: Wiley-Blackwell

Mackie, J.L. (1965) "Causes and Conditions," *American Philosophical Quarterly*, Vol. 2.4:245–264

MacWhinney, B. (2008) "How mental models encode embodied linguistic perspectives," in R. Klatzky, B. MacWhinney, & M. Behrmann (eds.), *Embodiment, Ego-Space, and Action*, pp. 369–410. Mahwah, NJ: Lawrence Erlbaum

Matthews, G. (1990) "Aristotelian Essentialism", *Philosophy and Phenomenomial Research*, Vol. L, Supplement 1990

McCawley, J. (1988) *The syntactic phenomena of English,* Chicago: The University of Chicago Press

MacNeilage, P. (2008) *The Origin of Speech*, Oxford: Oxford University Press (OUP)

Marty, A. (1918) *Untersuchungen zur Grundlegung der allgemeinen Grammatik und Sprachphilosophie*. Niemeyer, Halle an der Saale

McCawley, J.D. (1982a) "Parentheticals and Discontinuous Constituent Structure," *Linguistic Inquiry*, Vol. 13.1:91-106

McCawley, J.D. (1982b) *Thirty Million Theories of Grammar*, Chicago, Illinois: The University of Chicago Press

McCord, M.C. (1980) "Slot Grammars," *American Journal of Computational Linguistics*, Vol. 6.1:31–43

Mel'čuk, I. A. (1988) *Dependency Syntax: Theory and Practice*, Albany: State University of New York Press

Menon, V. (2011) "Developmental cognitive neuroscience of arithmetic: implications for learning and education," *ZDM (Zentralblatt für Didaktik der Mathematik)*, Vol. 42.6:515–525

Miller, G., and N. Chomsky (1963) "Finitary models of language users," in D. Luce (ed.), *Handbook of Mathematical Psychology*. John Wiley & Sons. pp. 2–419.

Montague, R. (1974) *Formal Philosophy*, Collected Papers, R. Thomason (ed.), New Haven: Yale University Press

Neisser, U. (1967) *Cognitive Psychology*, New York: Appleton-Century-Crofts

Neumann, J.v. (1945) *First Draft of a Report on the EDVAC*, in IEEE Annals of the History of Computing. Vol. 15, Issue 4, 1993, doi:10.1109/85.238389, pp. 27–75

Newell, A., and H.A. Simon (1972) *Human Problem Solving*, Englewood Cliffs, New Jersey: Prentice-Hall

Newton, I. ([1687, 1713, 1726], 1999) *Philosophiae Naturalis Principia Mathematicae* (The Principia: Mathematical Principles of Natural Philosophy), I.B. Cohen and A. Whitman (Translators), Berkeley: Univ. of California Press

Ogden, C.K., and I.A. Richards (1923) *The Meaning of Meaning*, London: Routledge and Kegan Paul

Osborne, T. (2006) "Gapping vs. non-gapping coordination," *Linguistische Berichte*, Vol. 207:307–338

Peirce, C.S. (1931–1935) *Collected Papers*. C. Hartshorne and P. Weiss (eds.), Cambridge, MA: Harvard Univ. Press

Persaud, K., and D. George (1982). "Analysis of discrimination mechanisms in the mammalian olfactory system using a model nose." *Nature* Vol. 299:352–5

Peters, S., and R. Ritchie (1973) "On the Generative Power of Transformational Grammar," *Information and Control*, Vol. 18:483–501

Post, E. (1936) "Finite Combinatory Processes — Formulation I," *JSL*, Vol. I:103–105

Post, E. (1946) "A variant of a recursively unsolvable problem," *Bull. Amer. Math. Soc.*, Vol. 52:264–269

Putten, S. van (2020) "Perception verbs and the conceptualization of the senses: The case of Avatime," *Linguistics: an International Review*, Vol. 58.2:425-462. doi: 10.1515/ling-2020-0039

Quillian, M. (1968) "Semantic memory," in M. Minsky (ed.), *Semantic Information Processing*, pp. 227–270, Cambridge, MA: MIT Press

Quine, W.v.O. (1960) *Word and Object*, Cambridge, Mass.: MIT Press

Read, S. (2017) "Aristotle's Theory of the Assertoric Syllogism," https://www.st-andrews.ac.uk/~slr/The_Syllogism.pdf

Reimer, M., and E. Michaelson (2014). "Reference," The Stanford Encyclopediaof Philosophy(Winter 2014 Edition), Edward N. Zalta (ed.), *http://plato.stanford.edu/archives/win2014/entries/reference/*

Rescher, N. (1969) *Many-valued Logic*, New York: McGraw-Hill

Rimé, B. (2009) "Emotion Elicits the Social Sharing of Emotion: Theory and Empirical Review," *Emotion Review*, Vol. 1.1:60–85

Rolls, E.T. (2000) "The Representation of Umami Taste in the Taste Cortex" *The Journal of Nutrition*, Vol.130.4:960–965

Robertson, I., T. and P. Atkins, "Essential vs. Accidental Properties", The Stanford Encyclopedia of Philosophy (Winter 2020 Edition), Edward N. Zalta (ed.), *https://plato.stanford.edu/archives/win2020/entries/essential-accidental*

Rosch, E. (1973) "Natural categories" *Cognitive Psychology*, Vol. 4.3:328–50, Amsterdam: Elsevier

Rosch, E. (1974) "Linguistic relativity" In A. Silverstein (Ed.), *Human communication: Theoretical perspectives*, New York: Halsted Press

Rosch, E. (1975) "Cognitive representations of semantic categories," *J. of Experimental Psychology*, General 104:192–253

Ross, J.R. (1967) *Constraints on Variables in Syntax*, Doctoral Dissertation, MIT (published as 'Infinite syntax!' Ablex, Norwood (1986))

Ross, J.R. (1970)"Gapping and the order of constituents," in M. Bierwisch and K. Heidolph (eds.), *Progress in linguistics: A collection of papers*, pp. 249–259, The Hague: Mouton

Ross, J.R. (1987) *Infinite Syntax!*, Norwood: Ablex

Sacks, H., E. Schegloff, and G. Jefferson (1974) "A Simplest Systematics for the Organization of Turn Taking for Conversation," *Language*, Vol. 50:696–735

Sag, I. (1976) *Deletion and Logical Form*, Doctoral Dissertation, MIT, Cambridge, Massachusetts

Saussure, F. de ([1916] 1972) *Cours de linguistique générale*, Édition critique préparée par Tullio de Mauro, Paris: Éditions Payot

Schegloff, E.A. (2007) *Sequence Organization in Interaction*, New York: Cambridge U. Press

Shannon, C.E., and W. Weaver (1949) *The Mathematical Theory of Communication*, Urbana, Illinois: University of Illinois Press

Skolem, T. (1920) "Logisch-kombinatorische Untersuchungen über die Erfüllbarkeit und Beweisbarkeit mathematischer Sätze nebst einem Theorem über dichte Mengen," Videnskapsselskapets Skrifter. I. Mat.-naturv. klasse, 1920. no. 4. Utgit for Fridtjof Nansens Fond

Slavov, M. (2013) "Newton's Law of Universal Gravitation and Hume's Conception of Causality," *Zeitschrift für philosophische Forschung*, Vol. 50.2:277–305

Snow, C.P. ([1959] 2001) *The Two Cultures*, London: Cambridge University Press (CUP)

Sperling, G. (1960) "The Information available in Brief Visual Processing," *Psychological Monographs*, 11, Whole No. 498

Spranger, M., M. Loetzsch, and S. Pauw (2010) "Open-ended Grounded Semantics" *https://csl.sony.fr/wp-content/themes/sony/uploads/pdf/spranger-10b.pdf*

Steels, L. (2008) "The symbol grounding problem has been solved. so what's next?" in *Symbols and Embodiment: Debates on Meaning and Cognition*, ed., M. de Vega, Oxford: Oxford University Press

Stubert, B. (1993) *"Einordnung der Familie der C-Sprachen zwischen die kontextfreien und die kontextsensitiven Sprachen,"* CLUE-betreute Studienarbeit der Informatik, Friedrich Alexander Universität Erlangen Nürnberg

Tarski, A. (1935) "Der Wahrheitsbegriff in den Formalisierten Sprachen," *Studia Philosophica*, Vol. I:262–405

Tarski, A. (1944) "The Semantic Concept of Truth," *Philosophy and Phenomenological Research*, Vol. 4:341–375

Tesnière, L. (1959) *Éléments de syntaxe structurale*, Paris: Editions Klincksieck

Tomita, M. (1985) "An efficient context-free parsing algorithm for natural languages," Proceedings of the *International Joint Conference on Artificial Intelligence*, pp. 756–764

Turing, A.M. (1950) "Computing machinery and intelligence," *Mind* 59:433–460

Venn, J. (1881) *Symbolic Logic*, London: MacMillan

Winquist, F. (2008) "Voltammetric electronic tongues - basic principles and applications," *Microchimica Acta*, Vol. 163.1-2:3–10

Wiriyathammabhum, P., D. Summers-Stay, C. Fermüller, and Y. Aloimonos (2016) "Computer Vision and Natural Language Processing: Recent Approaches in Multimedia and Robotics," December 2016, *ACM Computing Surveys*, Vol. 49(4):1-44

Wittgenstein, L. (1953) *Philosophische Untersuchungen*, Berlin: Suhrkamp Verlag

Younger, D. (1967) "Recognition and parsing of context-free languages in time n^3," *Information and Control*, Vol 10:2, pp. 189–208

Zadeh, L. (1965) "Fuzzy sets," *Information and Control*, Vol. 8:338–353

Zhang Yinsheng and Qiao Xiaodong (2009) "A Formal System of Aristotelian Syllogism Based on Automata Grammar," *Computer Science and Information Engineering*, 2009 WRI World Congress

Printed in the United States
by Baker & Taylor Publisher Services